Philosophy of Engineering and Technology

Volume 18

Editor-in-chief
Pieter E. Vermaas, Delft University of Technology, The Netherlands
General and overarching topics, design and analytic approaches

Editors
Christelle Didier, Lille Catholic University, France
Engineering ethics and science and technology studies
Craig Hanks, Texas State University, U.S.A.
Continental approaches, pragmatism, environmental philosophy, biotechnology
Byron Newberry, Baylor University, U.S.A.
Philosophy of engineering, engineering ethics and engineering education
Ibo van de Poel, Delft University of Technology, The Netherlands
Ethics of technology and engineering ethics

Editorial advisory board
Philip Brey, Twente University, the Netherlands
Louis Bucciarelli, Massachusetts Institute of Technology, U.S.A.
Michael Davis, Illinois Institute of Technology, U.S.A.
Paul Durbin, University of Delaware, U.S.A.
Andrew Feenberg, Simon Fraser University, Canada
Luciano Floridi, University of Hertfordshire & University of Oxford, U.K.
Jun Fudano, Kanazawa Institute of Technology, Japan
Sven Ove Hansson, Royal Institute of Technology, Sweden
Vincent F. Hendricks, University of Copenhagen, Denmark & Columbia University, U.S.A.
Don Ihde, Stony Brook University, U.S.A.
Billy V. Koen, University of Texas, U.S.A.
Peter Kroes, Delft University of Technology, the Netherlands
Sylvain Lavelle, ICAM-Polytechnicum, France
Michael Lynch, Cornell University, U.S.A.
Anthonie Meijers, Eindhoven University of Technology, the Netherlands
Sir Duncan Michael, Ove Arup Foundation, U.K.
Carl Mitcham, Colorado School of Mines, U.S.A.
Helen Nissenbaum, New York University, U.S.A.
Alfred Nordmann, Technische Universität Darmstadt, Germany
Joseph Pitt, Virginia Tech, U.S.A.
Daniel Sarewitz, Arizona State University, U.S.A.
Jon A. Schmidt, Burns & McDonnell, U.S.A.
Peter Simons, Trinity College Dublin, Ireland
Jeroen van den Hoven, Delft University of Technology, the Netherlands
John Weckert, Charles Sturt University, Australia

More information about this series at http://www.springer.com/series/8657

Sven Ove Hansson
Editor

The Role of Technology in Science: Philosophical Perspectives

 Springer

Editor
Sven Ove Hansson
Division of Philosophy
Royal Institute of Technology (KTH)
Stockholm, Sweden

ISSN 1879-7202   ISSN 1879-7210   (electronic)
Philosophy of Engineering and Technology
ISBN 978-94-017-9761-0   ISBN 978-94-017-9762-7   (eBook)
DOI 10.1007/978-94-017-9762-7

Library of Congress Control Number: 2015934901

Springer Dordrecht Heidelberg New York London
© Springer Science+Business Media Dordrecht 2015
This work is subject to copyright. All rights are reserved by the Publisher, whether the whole or part of the material is concerned, specifically the rights of translation, reprinting, reuse of illustrations, recitation, broadcasting, reproduction on microfilms or in any other physical way, and transmission or information storage and retrieval, electronic adaptation, computer software, or by similar or dissimilar methodology now known or hereafter developed.
The use of general descriptive names, registered names, trademarks, service marks, etc. in this publication does not imply, even in the absence of a specific statement, that such names are exempt from the relevant protective laws and regulations and therefore free for general use.
The publisher, the authors and the editors are safe to assume that the advice and information in this book are believed to be true and accurate at the date of publication. Neither the publisher nor the authors or the editors give a warranty, express or implied, with respect to the material contained herein or for any errors or omissions that may have been made.

Printed on acid-free paper

Springer Science+Business Media B.V. Dordrecht is part of Springer Science+Business Media (www.springer.com)

Preface

In some quarters philosophy seems to be conceived as an activity to be performed as independently as possible of the empirical conditions under which it takes place. The ideal philosopher, one would believe, is one who thinks as if he or she were a brain in a vat, capable of thinking on behalf of all possible, potential beings, independently of who we are. Needless to say there are no good reasons to believe this to be possible.

It is much more constructive to base our philosophical endeavours on the admittedly trivial but nevertheless important insight that philosophy is a human enterprise. It is an attempt to understand ourselves and the world that we live in. Therefore it must take into account the basic facts about how we as human beings interact with each other and with the world we live in. One of these basic facts is that we are tool-making animals. With the help of technology we have radically transformed – and continue to transform – the conditions under which we live and the ways in which we understand ourselves and the world.

In this perspective, technology has an important role to play in all branches of philosophical inquiry. The philosophy of science is one of the best examples. Few if any scientific investigations would be possible without technological devices. Many of our procedures for these investigations, such as experiments and measurements, have a strong technological background, and the same applies to important thought models that we employ in science, such as the notion of a mechanism. This book aims at bringing out the omnipresence of technology in science and showing why it must be closely attended to in philosophical reflections on science.

I would like to thank the publisher and the series editor Pieter Vermaas for their helpfulness and their strong support of this project and all the contributing authors for great work and for their commitment to this project.

Stockholm, Sweden
October 2, 2014

Sven Ove Hansson

Contents

Part I Introductory

1 **Preview** 3
   Sven Ove Hansson

2 **Science and Technology: What They Are and Why Their Relation Matters** 11
   Sven Ove Hansson

Part II The Technological Origins of Science

3 **Technological Thinking in Science** 27
   David F. Channell

4 **The Scientific Use of Technological Instruments** 55
   Mieke Boon

5 **Experiments Before Science. What Science Learned from Technological Experiments** 81
   Sven Ove Hansson

Part III Modern Technology Shapes Modern Science

6 **Iteration Unleashed. Computer Technology in Science** 113
   Johannes Lenhard

7 **Computer Simulations: A New Mode of Scientific Inquiry?** 131
   Stéphanie Ruphy

8 **Adopting a Technological Stance Toward the Living World. Promises, Pitfalls and Perils** 149
   Russell Powell

Part IV Reflections on a Complex Relationship

**9 Goal Rationality in Science and Technology.
An Epistemological Perspective** 175
Erik J. Olsson

**10 Reflections on Rational Goals in Science and Technology;
A Comment on Olsson** 193
Peter Kroes

**11 The Naturalness of the Naturalistic Fallacy and the Ethics
of Nanotechnology** 207
Mauro Dorato

**12 Human Well-Being, Nature and Technology.
A Comment on Dorato** 225
Ibo van de Poel

**13 Philosophy of Science and Philosophy of Technology:
One or Two Philosophies of One or Two Objects?** 235
Maarten Franssen

Contributors

Mieke Boon is Professor in Philosophy on a chair called Philosophy of Science in Practice in the Philosophy Department at the University of Twente. This department focuses on the philosophy of technology. Boon has a firm background in scientific research in the engineering sciences. She received a Ph.D. degree awarded cum laude in biotechnology at Technical University Delft. Her main philosophical interest concerns scientific research in (technological) application contexts. Between 2003 and 2008, she worked with a research grant from the Dutch National Science Foundation (NWO Vidi grant) on developing a philosophy of science for the engineering sciences. In 2012 she was awarded an Aspasia grant from the NWO. In 2006, she initiated the Society for Philosophy of Science in Practice (SPSP), of which she is a board member.

David F. Channell is Professor of the History of Ideas and Professor of Art and Technology at the University of Texas at Dallas. He received a B.S. in Physics from Case Institute of Technology, an M.S. in Physics from Case Western Reserve University, and a Ph.D. in the History of Science and Technology from Case Western Reserve University. He is the author of *Scottish Men of Science: William John Macquorn Rankine* (Scotland's Cultural Heritage 1986), *The History of Engineering Science: An Annotated Bibliography* (Garland 1989), and *The Vital Machine: A Study of Technology and Organic Life* (Oxford University Press 1991). He is currently completing a book entitled *Shifting Boundaries: How Engineering Became a Science and How Science Is Becoming Engineering*.

Mauro Dorato is Professor of Philosophy of Science at the Department of Philosophy of the Roma3 University. He is Director of the Ph.D. program and member of the Academy of Europe. He has been a member of the steering committee of the European Philosophy of Science Association, and president of the Italian Society for History and Philosophy of Science. Currently he is co-editor of the *European Journal for Philosophy of Science*. In addition to the philosophy of time, he has worked on scientific realism, philosophy of quantum mechanics, and laws of nature. On the latter topic, he has published *The Software of the Universe* (Ashgate 2005).

Maarten Franssen is Associate Professor at the Section of Philosophy and Ethics in the Faculty of Technology, Policy and Management of Delft University of Technology, the Netherlands. He teaches courses in philosophy and methodology of science and engineering, decision making, reasoning and argumentation in Delft and Leiden. His research interests and publications relate to design as rational decision making and its problems, normativity in relation to artefacts, the metaphysics of artefacts, and sociotechnical systems. He is co-author of *A Philosophy of Technology: From Technical Artefacts to Sociotechnical Systems* (Morgan and Claypool 2011) and co-editor of *Artefact Kinds: Ontology and the Human-Made World* (Springer 2014).

Sven Ove Hansson is Professor in Philosophy and Head of the Department of Philosophy and History, Royal Institute of Technology, Stockholm. He is member of the Royal Swedish Academy of Engineering Sciences (IVA) and was President of the Society for Philosophy and Technology in 2011–2013. He is editor-in-chief of *Theoria* and of the two book series *Outstanding Contributions to Logic* (Springer) and *Philosophy, Technology and Society* (Rowman & Littlefield). In addition to philosophy of science and technology he conducts research on logic, epistemology, decision theory, the philosophy of risk, and moral and political philosophy. His books include *A Textbook of Belief Dynamics* (Kluwer 1999), *The Structure of Values and Norms* (Cambridge University Press 2001), and *The Ethics of Risk* (Palgrave Macmillan 2013).

Peter Kroes is Professor in Philosophy of Technology at Delft University of Technology, the Netherlands. He has an engineering degree in physics (1974) and wrote a Ph.D. thesis on the notion of time in physical theories (University of Nijmegen, 1982). He has taught courses in the philosophy of science and technology and the ethics of technology, mainly for engineering students. His research in philosophy of technology focuses on technological artifacts and engineering design, socio-technical systems and technological knowledge. His most recent book publications are *Technical Artefacts: Creations of Mind and Matter* (Springer 2012), *A Philosophy of Technology: From Technical Artefacts to Socio-technical Systems* (together with Pieter Vermaas, Ibo van de Poel, Maarten Franssen and Wybo Houkes, Morgan and Claypool 2011) and *Functions in Biological and Artificial Worlds: Comparative Philosophical Perspectives* (edited with Ulrich Krohs, MIT Press 2009).

Johannes Lenhard does research in philosophy of science and engineering with a particular focus on the history and philosophy of mathematics and statistics. In recent years his research has focused on various aspects of computer and simulation modeling, culminating in his habilitation thesis *Calculated Surprises*. Currently, he is affiliated with Bielefeld University's Philosophy Department and the Center for Interdisciplinary Research (ZiF). He has held a visiting professorship in history at the University of South Carolina, Columbia, long after receiving his doctoral degree in mathematics from the University of Frankfurt. Asked for a sample paper, he names "Computer Simulation: The Cooperation Between Experimenting and Modeling", *Philosophy of Science*, 74 (2007), 176–194.

Erik J. Olsson is Professor and Chair in Theoretical Philosophy at Lund University, Sweden. His areas of research include epistemology, philosophical logic, pragmatism, and, more recently, philosophy of the Internet. He is associate editor of the journal *Theoria*, member of the Editorial Board of the book series *Studies in Epistemology* (Continuum), and co-founder and steering committee member of the European Epistemology Network. Olsson has contributed numerous book chapters and articles on subjects such as epistemic coherence, the value of knowledge, American pragmatism and social epistemology. Recent books include *Against Coherence: Truth, Probability, and Justification* (Oxford University Press 2005), *Knowledge and Inquiry: Essays on the Pragmatism of Isaac Levi* (Cambridge University Press 2006), and *Belief Revision Meets Philosophy of Science* (Springer 2011).

Russell Powell is Assistant Professor in the Department of Philosophy at Boston University. He has held faculty fellowships at the National Humanities Center, the American Council of Learned Societies, the Konrad Lorenz Institute for Evolution and Cognition Research, the Centre for Practical Ethics and the Institute for Science and Ethics at Oxford University, and the Kennedy Institute of Ethics at Georgetown University. Dr. Powell's research focuses primarily on conceptual, methodological and ethical problems in biological and biomedical science, especially in relation to evolutionary theory and emerging biotechnologies. Powell received a Ph.D. in Philosophy and M.S. in Evolutionary Biology from Duke University in 2009. Prior to commencing his graduate work in philosophy, he worked as an associate in the New York office of the global law firm Skadden, Arps, Slate, Meagher and Flom LLP, where he practiced complex pharmaceutical liability litigation. He is currently serving as associate editor for the *Journal of Medical Ethics*, which is part of the British Medical Journal Group.

Stéphanie Ruphy holds a Ph.D. in Astrophysics (Paris VI University) and a Ph.D. in Philosophy (Columbia University). She is currently Professor in the Philosophy of Science and head of the research laboratory Philosophie, Langages & Cognition at Grenoble-Alpes University in France. Her work in general philosophy of science has appeared in journals such as *Philosophy of Science, International Studies in the Philosophy of Science, Synthese,* and *Perspectives on Science*. She is also the author of *Pluralismes scientifiques. Enjeux épistémiques et métaphysiques* (Hermann 2013). Much of her work has concerned the unity or plurality of science debate, the role of values in science, and computer simulations.

Ibo van de Poel is Anthoni van Leeuwenhoek Professor in Ethics and Technology at Delft University of Technology. His research focuses on new technologies as social experiments, values in engineering design, moral responsibility, responsible innovation, engineering ethics, risk ethics, and the ethics of newly emerging technologies like nanotechnology. He is co-author of *Ethics, Engineering and Technology* (Wiley-Blackwell 2011), and co-editor of *Handbook of Philosophy of Technology and the Engineering Sciences* (Elsevier 2009), *Philosophy and Engineering* (Springer 2010), and *Moral Responsibility. Beyond Free Will and Determinism* (Springer 2011). He is also co-editor of the book series *Philosophy of Engineering and Technology* (Springer).

Part I
Introductory

Chapter 1
Preview

Sven Ove Hansson

Abstract This is a brief summary of the chapters in a multi-author book devoted to philosophical investigations of the role of technology in science. Some of the major themes treated in the book are: the role of technological devices, procedures and ways of thinking in science, how computer technology shapes modern science, goal rationality in science and technology, and the relations between technology and nature.

Technology is ubiquitous in science. Few scientific experiments or observations are performed without extensive use of technology. Computer simulations and other computational procedures are increasingly used in science, and concepts derived from technology are central in many scientific deliberations. This book investigates the many roles of technology in science and shows why they should be at the centre of attention in philosophical investigations on science.

In "Science and technology. What they are and why their relation matters" *Sven Ove Hansson* argues that a philosophical discussion of science-technology relationships has much to gain from clear definitions of the two key terms. The chapter takes us back to the medieval tradition of knowledge classification with its notions of science and mechanical arts, and shows how the modern concepts of science and technology have evolved from there. Both terms have acquired somewhat different meanings in different languages. The English language uses the word "science" in a limited sense that excludes the humanities, whereas the corresponding term in many other languages is broader and includes the humanities. It is proposed that the latter approach provides a more adequate delimitation from an epistemological point of view. The word "technology" originally referred to knowledge about practical activities with tools and machines, and this is still a common sense of the word for instance in German. In modern English, "technology" almost always refers to the tools, machines and activities themselves, rather than

---

S.O. Hansson (✉)
Division of Philosophy, Royal Institute of Technology (KTH), Brinellvägen 32,
10044 Stockholm, Sweden
e-mail: soh@kth.se

to knowledge about them. Based on these and other conceptual distinctions, the chapter ends by outlining four classes of philosophically interesting questions about science-technology relationships, namely those that refer to: (1) the relation between science in general and technological science, (2) the role of science in technological practice, (3) the role of technological practice in science, and (4) the relationship between science and the Aristotelian notion of productive arts (that is more general than the notion of technological practice).

1 The Technological Origins of Science

Historians of science have usually paid very limited attention to the influence of technology on science, and the little attention they have paid has usually referred to empirical methodologies employing new devices like the microscope and the spectrometer. In his chapter "Technological thinking in science" *David F. Channell* provides a broad overview of technology's impact on concepts and theories in science. Beginning in antiquity he shows how both Archimedes' and Hero's work included the use of technological models to solve what we would today call scientific problems. The invention of the mechanical clock in the late Middle Ages had a deep influence on scientific thinking that lasted well into the early modern period. The time meted out by a mechanical clock was uniform and independent of any terrestrial or celestial event, and this was the concept of time on which the new physical science could be built. Furthermore, the clock provided a model of a complex whole developing through the interaction of its parts, without any influence from the outside. This model served as the basis for new ideas both on the universe as a whole ("clockwork universe") and on biological creatures. The natural philosophies of Descartes, Gassendi, Hobbes, Boyle, Leibniz, and many others were deeply influenced by ideas based on the clock and other advanced mechanical devices. Technology was also an important inspiration for Francis Bacon's view that nature can be studied in artificial states, i.e. experimental set-ups. In the eighteenth and nineteenth centuries, technological developments were crucial for important scientific discoveries. Channell tells the story of how studies of the steam engine laid the foundations for the new science of thermodynamics, and also the somewhat less known story of how Maxwell and others used models from engineering science as starting-points in the development of electromagnetic theory. The chapter concludes with a discussion of the more recent phenomenon of an integrated "technoscience" that seemingly transcends the traditional divide between science and technology.

Channell's chapter is followed by two chapters exploring more specific ways in which technology has impacted on science. *Mieke Boon*'s chapter "The scientific use of technological instruments" has its emphasis on how natural science depends on technological instruments. This is a topic that has attracted some attention in recent philosophy of science, in particular in what Robert Ackerman has called New Experimentalism. By this is meant a recent tendency in the philosophy of science that is characterized by a focus on the role of experiments and instruments in science, and usually also by attempts to steer clear of the pitfalls of both logical empiricism and social constructivism. Boon provides an overview of the modern philosophical literature on technological instruments in science, covering issues such as the theory-ladenness of instrument-based observations, if and how we can know that we observe a natural phenomenon rather than an artefact of the experimental set-up, the use of instruments for discovery vs. hypothesis-testing, and the underdetermination of theory by empirical data. She also presents a typology that distinguishes between three roles of technological instruments in science, namely as Measure, Model, and Manufacture. By Measure is meant that the instrument measures, detects, or represents parameters of a natural object or process. A Model is a laboratory system that functions as a material model of a natural (or technological) object or process. Finally, by Manufacture is meant an apparatus that produces a phenomenon, typically one that is previously not known from nature but has been conjectured from theory. Concerning scientific practices in the engineering sciences, she emphasizes the crucial function of technological instruments in generating (or 'manufacturing') and investigating new physical phenomena that are of technological relevance. As most of philosophy of science is theory-oriented, the discipline tends to ignore this role of scientific instruments, thereby also suggesting that the contribution of science to technology is its theories. Boon emphasizes that a better understanding of the role of science in the development of advanced technologies requires taking into account the manufacturing role of technological instruments in the engineering sciences, that is, their role in scientific studies of technologically produced physical phenomena.

In "Experiments before science. What science learned from technological experiments" *Sven Ove Hansson* provides historical evidence that the first experiments were not scientific but instead directly action-guiding technological experiments. Systematic experimentation on agricultural and technological problems has taken place among indigenous people and craftspeople since long before the emergence of modern science. The purpose of these experiments was to achieve some desired practical result such as an improved harvest, a better mortar, glass or metal, or a slimmer but still sufficiently strong building. It is argued in this chapter that the philosophy of experimentation needs to draw a clear distinction between such "directly action-guiding" experiments and "epistemic" experiments that aim at understanding the workings of nature. Directly action-guiding experiments still have a major role for instance in technology and agriculture and (in the form of clinical trials) in medicine. It is argued that they differ from epistemic experiments in having a stronger and more immediate justification and in being less theory-dependent. However, the safeguards needed to avoid mistakes in execution and interpretation are essentially the same for the two types of experiments. Important such safeguards are control experiments, parameter variation, outcome measurement, blinding, randomization, and statistical evaluation. Several of these safeguards were developed by experimentalists working in pre-scientific technological traditions, and have been taken over by science for use also in epistemic experiments.

2 Modern Technology Shapes Modern Science

Few aspects of modern life are unaffected by the computer revolution. Science, for certain, has been affected at its very foundations. In "Iteration unleashed. Computer technology in science" *Johannes Lenhard* shows that the use of computer technology in science has philosophical implications, in particular by making new types of mathematical modeling possible. He focuses on two features of modern computer modeling. One is iteration that can be usefully exemplified by quantum chemical methods that are used to theoretically determine chemical structures and reactions. These methods provide a series of approximate solutions, each of which starts out with the previous solution and improves it. The calculation is finished when additional iterations no longer lead to significant changes. The other feature is exploration, by which is meant that a large number of inputs is tested, for instance in search of a minimum or maximum of some calculated variable. Monte Carlo methods are prominent versions of this exploratory approach. Lenhard proposes that from an epistemic point of view, computer modeling is characterized by epistemic opacity and agnosticism: opacity in the sense that the model's behaviour cannot be directly related to the input values, and agnosticism in the sense that computer models can provide conclusions that are based on resemblance of patterns rather than theoretical hypotheses. In this way, fundamental features of scientific thinking, such as our views of what constitutes an explanation or the basis for understanding a phenomenon, may be affected by the computer revolution in science.

Several themes from Lenhard's chapter are further discussed in *Stéphanie Ruphy*'s chapter "Computer simulations: a new mode of scientific inquiry?" By comparing simulations both to experimentation and theorizing, the two traditional paradigms of scientific activities, she attempts to clarify what is specific and new about computer simulations. She reviews most of the most discussed questions about simulations: Is the relationship between a computer simulation and its target system (that which we wish to know something about) the same as that between an experiment and its target system? Or is the similarity between a mathematical object and its (material) target system always weaker, and less supportive of inference, than the similarity between two material objects? Can a simulation be said to provide measurements, or should that term be reserved for procedures with a material component? (Some types of measurements, for instance in particle physics, are strongly model-dependent, so how important is the difference?) How do we determine the validity of a computer simulation? Can it provide us with explanations, or "only" with predictions? Ruphy proposes that to answer the last question we must distinguish between different types of computer simulations. Unfortunately there seems to be a trade-off between usefulness for explanatory and predictive purposes. The best simulation model for an explanatory purpose should be relatively simple, so that the workings of different submodels and the effects of changing various parameters are accessible to the user. The best simulation model for a predictive purpose may be the outcome of many efficient but unprincipled adjustments to make it yield the right answers. With such a model we may get the right answers (i.e. agreement with empirical observations) but for the wrong reasons (i.e. not because each of the submodels does exactly what is supposed to do.)

In his chapter "Adopting a technological stance toward the living world. Promises, pitfalls and perils" *Russell Powell* investigates the prevalence of technological thinking in a science where it is sometimes problematic, namely biology. In spite of the success of modern evolutionary theory organisms are often described and discussed both by laypeople and professional biologists as if they were designed artefacts. Why, he asks, does technological thinking still have such a prominent role in biology? A major reason seems to be that technological concepts are practically quite useful to describe and understand the outcome of evolutionary processes. The reason for this is of course that the same physical constraints apply both to evolutionary processes and deliberate design. Animal wings, developed independently in many different evolutionary lineages, have remarkable similarities both among themselves and with the wings of airplanes. The explanation is of course to be found in the laws of aerodynamics. The same engineering science can explain why streamlined and fusiform shapes reappear in many organisms and many technological artefacts. But having explained the usefulness of the "technological stance" in biology, Powell points out some of its pitfalls. The analogies can be extended beyond their domain of utility and prevent us from seeing important differences such as the self-organizing and self-reproducing nature of organisms and their developmental interactions with each other. In biological education, unqualified use of technological analogues may inadvertently lend support to creationism. Furthermore, in discussions about biologically based technology, language that equates organisms with technological artefacts can make us forget the ethically relevant differences between organisms and artefacts. Powell concludes that although the technological stance in biology has both cognitive and theoretical value, it has serious perils and should therefore not be adopted without qualification.

3 Reflections on a Complex Relationship

Science and technology are conventionally distinguished in terms of their goals. The goal of science is truth; that of technology is practical usefulness. In his chapter "Goal rationality in science and technology. An epistemological perspective" *Erik J. Olsson* argues that although science and technology have different goals, their criteria of goal-setting rationality may be the same. He shows this by applying to science a set of rationality criteria for goal-setting that has been developed for non-epistemic goals such as those of management and technological design. In doing this he works through four epistemological debates, all of which concern goal rationality in science: Peirce's argument that the goal of scientific inquiry should not be truth or true belief, but merely belief or opinion, Rorty's somewhat related claim that truth should be replaced as the goal of inquiry by justified belief, Kaplan's assertion that knowledge is not an adequate goal of inquiry, and Sartwell's contention that knowledge is nothing else than mere true belief. Olsson endeavours to show that in all four cases, conclusions have been drawn on the basis of standards for rational goal-setting that do not appear to be valid. The most common problem seems to be that no attention has been paid to the motivating role of goals. According to Olsson, pragmatist philosophers have been rash when denouncing truth as a goal of inquiry. Instead "the goal of truth should rather be cherished by pragmatists as a goal which, due to its tendency to move inquirers to increase their mental effort, is as practice-affecting as one could have wished". Thus, interestingly, an application of technology-style goal rationality to science may lead to a reinforcement of the traditional goal of scientific inquiry.

Peter Kroes's commentary, "Reflections on rational goals in science and technology. A comment on Olsson", has two main parts. In the first part he criticizes Olsson's approach to the goal of technology. According to Kroes, technology's goal is to make useful things, and this is in itself not a knowledge goal (although knowledge can be a means to achieve it). Usefulness is context-dependent in a way that knowledge is not, and therefore technology is also context-dependent in a way that science is not. In the second part, Kroes expresses doubts on the theory of rational goal-setting that Olsson employs and raises the question how this notion of rationality relates to the traditional (Humean) view according to which practical rationality is nothing else than instrumental rationality that takes goals for given. In such a view, goal-setting rationality is an oxymoronic notion. According to Kroes, the criteria for rational goal-setting that Olsson refers to fall within the domain of instrumental rationality, and to the extent that they are useful this does not contradict the standard view that rationality is always concerned with means rather than ends. This conclusion tends to undermine Olsson's proposal that these criteria can usefully be applied to the philosophical issue what is the most appropriate overarching goal of science.

In his chapter "The naturalness of the naturalistic fallacy and the ethics of nanotechnology" *Mauro Dorato* investigates the ethical aspects of the technology–nature dichotomy. His ethical viewpoint is based on the Aristotelian notion of human flourishing, and from that stance he rejects the common view that the presumed "unnaturalness" of new technologies has ethical implications. Instead of focusing on the contrast natural–artificial we should focus on finding out what is beneficial or harmful for human flourishing. This may seem self-evident, so why is the idealization of nature such a common tendency? Dorato proposes that a preference for stability in the natural environment may have had survival value in previous phases of human history. However, in our present situation, a general rejection of technologies initially seen as "unnatural" would lead us seriously wrong. Dorato uses microscopic bodily implants as examples. Such implants can potentially be used for eminently useful medical purposes. For instance, brain implants can be developed that detect an approaching epileptic seizure on the pre-ictal stage and prevent its onset. Other microimplants can be used to administer personalized and localized pharmacological treatments. Dorato concludes by defending the use of these and other advanced technologies. Although they initially seem to be "against nature", they bring more promise than menace, and therefore we will gain from learning to live with them.

Ibo van de Poel's commentary on Dorato's text, "Human well-being, nature and technology", focuses on two major issues. First, he considers Dorato's claim that taking nature as a norm had survival value for previous generations but does not have so any longer. According to van de Poel, it may still have some survival value. For our species to survive, we need to put certain limits on our interventions in the environment. Therefore, some nature-based norms may be useful, although they need not coincide with the traditional idealization of naturalness. Secondly, van de Poel questions Dorato's use of human flourishing as a primary ethical criterion. The concept of flourishing is too vague, he says, and persons with contradictory views in a moral issue can all claim that their view is conducive to human flourishing. Instead, van de Poel proposes that the primary criterion for the evaluation of technologies should be human well-being, supplemented by other values such as justice and sustainability.

In the concluding chapter, "Philosophy of science and philosophy of technology: one or two philosophies of one or two objects?" *Maarten Franssen* discusses the science-technology relationship in a more general perspective. In practice, science and technology are so tightly interwoven that some people prefer to speak of "technoscience" as a unified entity. But philosophically, science and technology are far apart. One reason for this, says Franssen, may be the historical differences between the two philosophical disciplines. The philosophy of science is reasonably focused, and deals with the methods of science and the epistemic justification of its outputs. The philosophy of technology is much wider in scope and deals with the relations of technology to culture, society, and the "essence of mankind". Corresponding approaches to science have never been included in the philosophy of science. But apart from that, is it possible to distinguish philosophically between science and technology? Franssen believes it is, although this will have to be a "surgical dissection, so to speak, as if separating a pair of conjoined twins". The only reasonable placement of the cut, he says, would have to be between an ideal-typical pure science with theoretical rationality as its guiding principle and an ideal-typical pure technology whose guiding principle is practical rationality. However, neither of these is easily delimited. Franssen mentions several problems for the delimitation. One of them is that the "purely theoretical" notion of science has been seriously questioned. If we adopt for instance the instrumentalist view that science aims at accounting for the world, rather than telling us the truth about it, then belief formation can be seen as an implementation of practical rather than theoretical rationality. This would bring science and technology closer to each other in a philosophical sense. Franssen concludes by proposing that perhaps the distinction between science and technology does not serve us well, and then the same would apply to that between philosophy of science and philosophy of technology. Perhaps, he says, we had better replace the two philosophical subdisciplines, currently operating at a vast distance from each other, by a single philosophy of technoscience.

Chapter 2
Science and Technology: What They Are and Why Their Relation Matters

Sven Ove Hansson

Abstract The relationship(s) between science and technology can be conceived in different ways depending on how each of the two concepts is defined. This chapter traces them both back to the medieval tradition of knowledge classification and its notions of science and mechanical arts. Science can be defined either in the limited sense of the English language or in a broader sense that includes the humanities. It is argued that the latter approach provides a more adequate delimitation from an epistemological point of view. The word "technology" can refer either to knowledge about practical activities with tools and machines (a common sense in German and many other languages) or to these activities, tools, and machines themselves (the common sense of the word in English). Based on conceptual clarifications of the two concepts, four classes of philosophically interesting questions about science-technology relationships are outlined: (1) the relation between science in general and technological science, (2) the role of science in technological practice, (3) the role of technological practice in science, and (4) the relationship between science and the Aristotelian notion of productive arts (that is more general than the notion of technological practice).

1 Introduction

Before delving into the relationship(s) between science and technology we should pay some attention to the meanings of each of these two terms. Do they represent important and well-demarcated concepts, or are they delimited in unsystematic ways that make them unsuitable as objects of philosophical reflection? We will begin by tracing their origins in the classifications of knowledge that had a prominent role in academic treatises from the Middle Ages and well into the modern age. Section 2 introduces the medieval tradition of knowledge classification, and Sect. 3 the place of what we now call technology in these classifications systems. Sections 4 and 5 discuss the origins and the vagaries of the terms "science" respectively

---

S.O. Hansson (✉)
Division of Philosophy, Royal Institute of Technology (KTH), Brinellvägen 32, 10044 Stockholm, Sweden
e-mail: soh@kth.se

"technology". In Sect. 6 it is suggested that attention to the different meanings of the two terms can help us to distinguish in a more precise way between different approaches to what we call the "science – technology relationship".

2 Knowledge Classification

The classification of areas of human knowledge was a recurrent theme in learned expositions throughout the Middle Ages. A large number of classification schemes have survived, usually with a tree-like structure that organized the various disciplines in groups and subgroups. These classification schemes[1] served to identify the areas worthy of scholarly efforts, and often also to list the disciplines to be included in curricula (Dyer 2007; Ovitt 1983). But despite the great care that was taken in listing and categorizing the different branches of knowledge, not much importance seems to have been attached to the choice of a general term to cover all knowledge. "Scientia" (science), "philosophia" (philosophy), and "ars" (arts) were all used for that purpose.

Etymologically, one might expect a clear distinction between the three terms. "Scientia" is derived from the verb "scire" (to know) that was used primarily about knowledge of facts. "Philosophia" is a Greek term that literally means "love of wisdom", but it was often interpreted as systematic knowledge and understanding in general, both about facts and about more speculative topics such as existence and morality. Cicero influentially defined it as follows:

> [P]hilosophy is nothing else, if one will translate the word into our idiom, than 'the love of wisdom'. Wisdom, moreover, as the word has been defined by the philosophers of old, is 'the knowledge of things human and divine and of the causes by which those things are controlled.'[2] (Cicero, *De Officiis* 2.5)

"Ars" refers to skills, abilities, and craftsmanship. It was the standard translation of the Greek "techne". Aristotle provided an influential and ingenious definition of the concept that has often been referred to as a definition of the productive arts:

> Now since architecture is an art and is essentially a reasoned state of capacity to make, and there is neither any art that is not such a state nor any such state that is not an art, art is identical with a state of capacity to make, involving a true course of reasoning. All art is concerned with coming into being, i.e. with contriving and considering how something may come into being which is capable of either being or not being, and whose origin is in the maker and not in the thing made; for art is concerned neither with things that are, or come into being, by necessity, nor with things that do so in accordance with nature (since these have their origin in themselves). (Aristotle, *Nichomachean Ethics* VI:4)

---

[1] *Divisiones scientiarum* or *divisiones philosophiae*.

[2] [N]ec quicquam aliud est philosophia, si interpretari velis, praeter studium sapientiae. Sapientia autem est, ut a veteribus philosophis definitum est, rerum divinarum et humanarum causarumque, quibus eae res continentur, scientia.

But in spite of their differences in meaning, all three terms were used interchangeably as umbrella terms for all knowledge. The usage differed between authors in what seems to be a very unsystematic way. Some authors used "science" as the most general term and "philosophy" as a second-level term to denote some broad category of knowledge disciplines. Others did exactly the other way around, and still others used "science" and "philosophy" as synonyms. Similarly, "art" was sometimes used to cover all the disciplines, sometimes to cover some broad subcategory of them. This terminological confusion persisted well into the sixteenth and seventeenth centuries (Covington 2005; Freedman 1994; Ovitt 1983). For a modern reader it may be particularly surprising to find that in the Middle Ages, "philosophy" included all kinds of knowledge, also practical craftsmanship. From the end of the fifteenth century it became common to exclude the crafts (the mechanical arts) from philosophy, but as late as in the eighteenth century the word "philosophy" was commonly used to denote all kinds of knowledge (Freedman 1994; Tonelli 1975).

3 The Mechanical Arts

In medieval and early modern times, the term "art" (ars) referred to all kinds of skills and abilities. It did not suggest a connection with what we today call the "fine arts" or just "art". The notion of art included "not only the works of artists but also those of artisans and scholars" (Tatarkiewicz 1963, 231).[3] The arts emphasized in knowledge classifications were the so-called "liberal arts". This is a term used since classical antiquity for the non-religious disciplines usually taught in schools, so called since they were the arts suitable for free men (Chenu 1940; Tatarkiewicz 1963, 233). Medieval universities had four faculties: Theology, Law, Medicine, and the Arts. The former three were the higher faculties to which a student could only be admitted after studying the liberal arts at the Faculty of Arts (also called the Faculty of Philosophy) (Kibre 1984).

Since the early Middle Ages, the liberal arts were usually considered to be seven in number, and divided into two groups. A group of three, called the "trivium" consisted of what we may call the "language-related" disciplines, namely logic, rhetoric, and grammar. The other group, the "quadrivium", consisted of four mathematics-related subjects, namely arithmetic, geometry, astronomy, and

---

[3] It was not until the eighteenth century that a literature emerged in which the fine arts were compared to each other and discussed on the basis of common principles. The term "fine arts" (in French "beaux arts") was introduced to denote painting, sculpture, architecture, music, and poetry, and sometimes others artforms such as gardening, opera, theatre, and prose literature. The decisive step in forming the modern concept of art was taken by Charles Batteux (1713–1780), professor of philosophy in Paris. In his book from 1746, *Les beaux arts réduits à un même principe* (The fine arts reduced to a single principle), he for the first time clearly separated the fine arts such as music, poetry, painting, and dance from the mechanical arts (Kristeller 1980).

music. By music was meant a theoretical doctrine of harmony that had more in common with mathematics than with musicianship (Dyer 2007; Freedman 1994; Hoppe 2011; James 1995). Various authors made additions to the list of liberal arts, claiming that one or other additional activity should be counted as a liberal art. Not surprisingly, Vitruvius saw architecture as a liberal art, and Galen wanted to add medicine to the list. Others wanted to give agriculture that status, probably due to its association with a simple, innocent life (Van Den Hoven 1996).

The liberal arts explicitly excluded most of the activities undertaken for a living by the lower and middle classes. In antiquity such arts were called illiberal, vulgar, sordid, or banausic.[4] These were all derogative terms, indicating the inferior social status of these activities and reflecting a contemptuous view of physical work that was predominant in classical Greece (Van Den Hoven 1996, 90–91; Ovitt 1983; Tatarkiewicz 1963; Whitney 1990). In the Middle Ages, the most common term was "mechanical arts".[5] It was introduced in the ninth century by Johannes Scotus Eriugena in his commentary on Martianus Capella's allegorical text on the liberal arts, *On the Marriage of Philology and Mercury*.[6] According to Johannes Scotus, Mercury gave the seven liberal arts to his bride, Philology, and in exchange she gave him the seven mechanical arts. However, Scotus did not name the mechanical arts (Van Den Hoven 1996; Whitney 1990). Instead a list of seven mechanical arts, or rather groups of arts, was provided in the late 1120s by Hugh of Saint Victor:

1. lanificium: weaving, tailoring;
2. armatura: masonry, architecture, warfare;
3. navigatio: trade on water and land;
4. agricultura: agriculture, horticulture, cooking;
5. venatio: hunting, food production;
6. medicina: medicine and pharmacy;
7. theatrica: knights' tournaments and games, theater. (Hoppe 2011, 40–41)

The reason why Hugh summarized the large number of practical arts under only seven headings was obviously that he desired a parallel with the seven liberal arts. Hugh emphasized that just like the liberal arts, the mechanical ones could contribute to wisdom and blessedness. He also elevated their status by making the mechanical arts one of four major parts of philosophy (the others being theoretical, practical, and logical knowledge) (Weisheipl 1965, 65). After Hugh it became common (but far from universal) to include the mechanical arts in classifications of knowledge (Dyer 2007).

The distinction between liberal and mechanical arts continued to be used in the early modern era, and it had an important role in the great French *Encyclopédie*, published from 1751 to 1772, that was the most influential literary output of the Enlightenment. One of its achievements was the incorporation of the mechanical

---

[4] Artes illiberales, artes vulgares, artes sordidae, artes banausicae.

[5] Artes mechanicae.

[6] De nuptiis Philologiae et Mercurii.

arts, i.e. what we call technology, into the edifice of learning. In the preface Jean Le Rond d'Alembert (1717–1783) emphasized that the mechanical arts were no less worthy pursuits than the liberal ones.

> The mechanical arts, which are dependent upon manual operation and are subjugated (if I may be permitted this term) to a sort of routine, have been left to those among men whom prejudices have placed in the lowest class. Poverty has forced these men to turn to such work more often than taste and genius have attracted them to it. Subsequently it became a reason for holding them in contempt – so much does poverty harm everything that accompanies it. With regard to the free operations of the mind, they have been apportioned to those who have believed themselves most favoured by Nature in this respect. However, the advantage that the liberal arts have over the mechanical arts, because of their demands upon the intellect and because of the difficulty of excelling in them, is sufficiently counterbalanced by the quite superior usefulness which the latter for the most part have for us. It is this very utility which has reduced them forcibly to purely mechanical operations, so that the practice of them may be made easier for a large number of men. But society, while rightly respecting the great geniuses which enlighten it, should in no wise debase the hands which serve it. (d'Alembert 1751, xiij)

4 The Modern Term "Science"

The English word "science" derives from the Latin "scientia", and originally, it had an equally wide meaning. It could refer to almost anything that you had to learn in order to master it: everything from scholarly learning to sewing and horse riding. But in the seventeenth and eighteenth centuries the meaning of "science" was restricted to systematic knowledge. The word could for instance refer to the knowledge you need to make a living in a particular practical trade. In the nineteenth century the meaning of "science" was further restricted, and it essentially meant what we would today call natural science (Layton 1976). Today, the term "science" is still primarily used about the natural sciences and other fields of research that are considered to be similar to them. Hence, political economy and sociology are counted as sciences, whereas literature and history are usually not. In several academic areas considerable efforts have been devoted to making one's own discipline accepted as a science. This applies for instance to social anthropology that is often counted as a science although it is in many respects closer to the humanities (Salmon 2003).

Thus, given the current meaning of the term, far from all knowledge can be described as scientific. However, the distinction between scientific and non-scientific knowledge depends not only on epistemological principles but also on historical contingencies. This we can see clearly from the difference in meaning between the word "science" in English and the corresponding word "Wissenschaft" in German with its close analogues in Dutch and the Nordic languages. "Wissenschaft" also originally meant knowledge, but it has a much broader meaning than "science". It includes all the academic specialties, including the humanities. With its wider area of application, "Wissenschaft" is closer than "science" to "scientia".

In my view, the German term "Wissenschaft" has the advantage of giving a more adequate delimitation from an epistemological point of view than the English term. "Wissenschaft" does not exclude academic or otherwise systematized knowledge disciplines such as history and other humanities that are excluded from the "sciences" due to linguistic conventions. The restricted sense of the English word "science" is unfortunate since the sciences and the humanities share a common ground, in at least two respects. First, their very raison d'être is the same, namely to provide us with the most epistemically warranted statements that can be made, at the time being, on the subject matter within their respective domains.

Secondly, they are intricately connected, and together they form a *community of knowledge disciplines* that is characterized and set apart by mutual respect for each other's results and methods (Hansson 2007b). Such mutual respect is something that we take for granted for instance between physics and chemistry, but it also holds across the (contrived) boundary between the sciences and the humanities. An archaeologist or historian will have to accept the outcome of a state-of-the art chemical analysis of an archaeological artefact. In the same way, a zoologist will have to accept the historians' judgments of the reliability of an ancient text describing extinct animals. In order to understand ancient descriptions of diseases we need co-operations between classical scholars and medical scientists (and most certainly not between classical scholars and homeopaths or between medical scientists and bibliomancers).

Neither "science" nor any other established term in the English language covers all the members of this community of knowledge disciplines. For lack of a better term, I will call them "science(s) in a broad sense". The name is not important, but it is important to recognize that we have a community of knowledge disciplines that all strive to obtain reliable knowledge and all respect the other disciplines in their respective areas of speciality. Many discussions on science (such as that about the science–pseudoscience distinction) seem to refer in practice to science in the broad sense, but that is not always made as clear as it should be (Hansson 2013b).

Science, in this broad sense, is an epistemological, not a sociological category. The knowledge disciplines belonging to science in the broad sense are characterized by a common aim, namely to provide us with the most epistemically warranted information that can be obtained in subject-matter within their respective domains. This definition is close to coinciding with the academic disciplines, but it does not coincide exactly with them. There are some (minor) branches of learning that satisfy the inclusion criteria but do not have academic status. This applies for instance to philately and to the history of conjuring, both of which are pursued by devoted amateurs rather than by professional scholars.

5 The Modern Term "Technology"

The word "technology" is of Greek origin, based on "techne" that means art or skill and "-logy" that means "knowledge of" or "discipline of". The word was introduced into Latin as a loanword by Cicero (Steele 1900, 389).[7] However, it

---

[7]Cicero, *Epistulae ad Atticum* 4:16.

does not seem to have been much used until Peter Ramus (1515–1572) started to use it in the sense of knowledge about the relations among all technai (arts). The word became used increasingly to denote knowledge about the arts. In 1829 the American physician and scientist Jacob Bigelow published *Elements of Technology* where he defined technology as "the principles, processes, and nomenclatures of the more conspicuous arts, particularly those which involve applications of science" (Tulley 2008). Already in the late seventeenth century "technology" often referred specifically to the mechanical arts and the skills of craftspeople (Sebestik 1983). This sense became more and more dominant, and in 1909 *Webster's Second New International Dictionary* defined technology as "the science or systematic knowledge of industrial arts, especially of the more important manufactures, as spinning, weaving, metallurgy, etc." (Tulley 2008). This means that technology was no longer conceived as knowledge about *techne* in the original Greek sense of the term, i.e. arts and skills in general. It had acquired a more limited sense referring to what is done with tools and machines.

This delimitation of techne and technology excludes many skills (or "productive arts"). We do not usually use the term "technology" to refer to knowledge about the skills of a physician, a cook, or a musician. On the other hand we tend to use the term about computer programming and software engineering. The delimitation of skills counted as technological appears rather arbitrary, in much the same way as the exclusion of history and art theory from science appears arbitrary. Arguably, the Aristotelian sense of "ars" (or "techne") is more principled and coherent than the modern delimitation of "technology".

But in the English language the word "technology" also acquired another meaning that became more and more common: Increasingly it referred to the tools, machines, and procedures used to produce material things, rather than to science or knowledge about these tools, machines, and procedures. This usage seems to have become common only in the twentieth century. The earliest example given in the Oxford English Dictionary is a text from 1898 about the coal-oil industry, according to which "a number of patents were granted for improvements in this technology, mainly for improved methods of distillation" (Peckham 1898, 119). Today this is the dominant usage. As Joost Mertens noted, "[i]n English usage, 'technology' normally refers to instrumental practices or their rules and only exceptionally to the scientific description, explication or explanation of these practices." (Mertens 2002). However, this is not true of all languages. For instance, French, German, Dutch, and Swedish all have a shorter word (technique, Technik, techniek, teknik) that refers to the actual tools, machines and practices. In these languages, the word corresponding to "technology" (technologie, Technologie, technologie, teknologi) is more often than in English used to denote knowledge about these practical arts rather than to denote these arts and their material devices themselves. However, due to influence from English, the use of "technology" in the sense of tools, machines and practices is common in these languages as well. (According to the *Svenska Akademiens Ordbok*, the Swedish counterpart of the OED, this usage seems to have become common in Swedish in the 1960s.)

6   Interrelations Between Science and Technology

Given all these meanings of "science" and "technology", we can mean different things when discussing the relationship between science and technology. As to science, the crucial difference is that between the restricted sense of the word in modern English and the broader sense attached both to its Latin ancestor "scientia" and to the corresponding words in German and several other languages. From an epistemological point of view, the broader sense is more interesting since, as I noted above, it represents a more principled, less arbitrary demarcation. From a sociological point of view, on the other hand, there may be good reasons to focus on "science" in the conventional English sense of the word. Arguably science (in this sense) has a social identity or role not shared by the humanities; not least in relation to engineering and technology (in at least some senses of the latter word).

Turning to technology, there are even more options. First of all, we must distinguish between technology as systematic knowledge about practices involving tools, machines etc., and technology as these practices themselves. We can call the first of these technology-knowledge and the second technology-practice.[8] The relationship between technology-knowledge and science would seem to be one of subsumption rather than conflict. In other words, technology-knowledge is a branch of science rather than something that runs into conflict with science. But as already mentioned, this is not the common sense of "technology" in English. To refer to this concept in English it is probably best to use the phrase "technological science".

Technology-practice is a subclass of the "productive arts" in the Aristotelian sense, since it is concerned with the creation of something new. It consists mainly of those productive arts that produce material things with the help of tools or machines. There are also other productive arts that we do not usually call "technology", such as the arts of medicine, farming, music, dance, etc. We seldom use the phrase "productive arts" today, but that does not make the category philosophically uninteresting. It is reasonable to ask whether some of the philosophical issues that we discuss in relation to technology can be generalized in an interesting way to the productive arts.

Some of these distinctions are summarized in Fig. 2.1. The left circle represents technological science (technology-knowledge), whereas the right circle represents the most common meaning of "technology" in English, namely technology-practice. The ellipse surrounding technological science represents the wider category of science in general (taken here preferably in the broad sense), whereas that surrounding technology-practice represents the useful but today largely obliviated Aristotelian concept of the productive arts. Given these conceptual clarifications, there are at least four classes of interesting philosophical problems about the relationships between science and technology. They are schematically represented in Fig. 2.2.

---

[8]Or techno*logy* and *techno*logy.

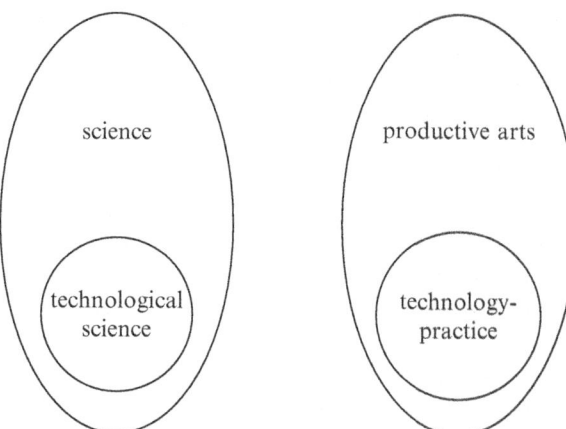

Fig. 2.1 Two major meanings of "technology" are technological science, that makes it a subcategory of science, and technology-practice that makes it a subcategory of the productive arts

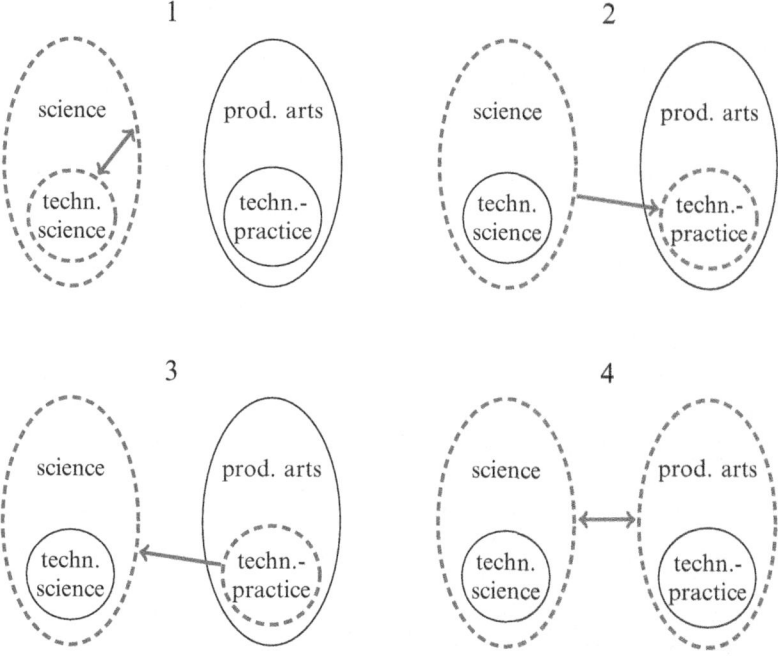

Fig. 2.2 Four philosophically interesting explications of the notion of a science-technology relationship: (1) the relation between science in general and technological science, (2) the role of science in technological practice, (3) the role of technological practice in science, and (4) the relationship between science and the Aristotelian notion of productive arts (that is more general than the notion of technological practice)

First, we have questions about the relationship between science and technology-knowledge, or between science in general and technological science. When discussing this, we can mean by "science in general" either science in the restricted English-language sense that excludes the humanities or in a broader sense that includes them. One important research question is whether the technological sciences differ from other sciences in other respects than their subject matter, for instance whether they have different methodologies or epistemological criteria.[9] Another such question is whether the technological sciences are applied natural sciences, i.e. entirely based on principles referring to objects that are not human-made, or whether additional principles are needed that refer to the human creation of technological artefacts.

Secondly, we have questions about the role of science (in either the conventional or the broad sense) in technology-practice. To what extent is technological practice, such as various forms of engineering, based on scientific knowledge? Today it is commonplace that technology-practice is not just applied science. It also involves other types of knowledge, such as tacit knowledge and (explicit but non-scientific) rules of thumb. What is the nature of such knowledge, and how does it differ from scientific knowledge? (Hansson 2013a; Norström 2011) (This second group of questions should be distinguished from the first group that refers to the relationship between science in general and technological science.)

Thirdly, there are interesting questions about the reverse relationship, namely the role of technology-practice in science (in either the conventional or the broad sense). The Austrian historian and philosopher Edgar Zilsel (1891–1944) showed that Galileo Galilei (1564–1642) and other scientific pioneers depended on the help of skilled workers in order to succeed in extracting information from nature by manipulating it, i.e. making experiments (Drake 1978; Zilsel 1942, 2000). In more recent years, several authors have claimed that it is more accurate to describe

---

[9]The following six differences between technological and natural science were proposed in (Hansson 2007a).

1. Their primary study objects have been constructed by humans, rather than being objects from nature.
2. Design is an important part of technological science. Technological scientists do not only study human-made objects, they also construct them.
3. The study objects are largely defined in functional, rather than physical, terms.
4. The conceptual apparatus of the technological sciences contains a large number of value-laden notions. (Examples are 'user friendly', 'environmentally friendly', and 'risk'.)
5. There is less room than in the natural sciences for idealizations. For instance, physical experiments are often performed in vacuum in order to correspond to theoretical models in which the impact of atmospheric pressure has been excluded, and for similar reasons chemical experiments are often performed in gas phase. In the technological sciences, such idealizations cannot be used.
6. In mathematical work, technological scientists are satisfied by sufficiently good approximations. In the natural sciences, an analytical solution is always preferred if at all obtainable.

science as applied technology than the other way around (Lelas 1993). The use of technology in science is at focus in most of the chapters that follow.

Fourthly, we can generalize these deliberations to an arguably more philosophically fundamental level, namely the relationship between on the one hand science in the broad sense and on the other hand the productive arts, or goal-directed practical activities, in general.[10] This will in fact be a resumption of the way in which the relationship between science and the arts was studied long before the modern humanities-excluding notion of science, and long before the modern notion of technology that only includes a fraction of the practical arts. The English philosopher Robert Kilwardby (1215–1279) discussed this relationship in a remarkably sophisticated way. He emphasized that a distinction must be made between science in a broad sense (called "speculative philosophy") and the practical skills, but he also pointed out that they are dependent on each other in a fundamental way:

> In as much as we have said something separately concerning the speculative part of philosophy and something about the practical part, now it is important to say something about them in comparison with each other. I ask therefore in what way they are distinguished according to their degree of speculative philosophy and praxis, since those which are practical are, indeed, speculative – it is important certainly that one consider first by speculative virtue what one ought to perform in practical virtue – and, conversely, the speculative sciences are not without praxis. Does not, in fact, arithmetic teach how to add numbers to each other and to subtract them from each other, to multiply and divide and draw out their square roots, all of which things are operations? Again does not music teach to play the lute and flute and things of this sort? Again does not geometry teach how to measure every dimension, through which both carpenters and stoneworkers work? Again, does not one know the time for navigation and planting and things of this sort through astronomy? It seems therefore that every single science said to be speculative is also practical. It seems, therefore, that the speculative sciences are practical and the practical speculative. (Quoted from Whitney 1990, 120)[11]

Seen in this wider perspective, elucidation of the science-technology relationships is important not only for the philosophy of science and the philosophy of technology, but also more broadly for our philosophical understanding of the relationships between human knowledge and human activity, and between theoretical and practical rationality.

---

[10]In a similar vein, the German historian of technology Otto Mayr has proposed a research focus on "historical interactions and interchanges between what can roughly be labeled 'theoretical' and 'practical' activities, that is, between man's investigations of the laws of nature and his actions and constructions aimed at solving life's material problems." (Mayr 1976, 669).

[11]In his *Opera Logica* (1578) the Italian philosopher Jacopo Zabarella (1533–1589) discussed the same issue, but reached a different conclusion. In his view, the productive arts can learn from science but not the other way around (Mikkeli 1997, 222).

References

Aristotle. (1980). *Nichomachean ethics* (W. D. Ross, Trans.). Oxford: Clarendon Press.
Chenu, M. -D. (1940). Arts 'mecaniques' et oeuvres serviles. *Revue des sciences philosophiques et theologiques, 29*, 313–315.
Cicero, M. T. (1913). *De officiis* (Loeb classical library, W. Miller, English Trans.). Cambridge: Harvard University Press.
Covington, M. A. (2005). Scientia sermocinalis: Grammar in medieval classifications of the sciences. In N. McLelland & A. Linn (Eds.), *Flores grammaticae: Essays in memory of Vivien Law* (pp. 49–54). Münster: Nodus Publikationen.
d'Alembert, J. le R. (1751). *Discours préliminaire*. In D. Diderot & J. le R. d'Alembert (Eds.), *Encyclopédie, ou dictionnaire raisonné des sciences, des arts et des métiers, par une société de gens de lettres* (Vol. 1). Paris: Briasson.
Drake, S. (1978). *Galileo at work: His scientific biography*. Chicago: University of Chicago Press.
Dyer, J. (2007). The place of musica in medieval classifications of knowledge. *Journal of Musicology, 24*, 3–71.
Freedman, J. S. (1994). Classifications of philosophy, the sciences, and the arts in sixteenth- and seventeenth-century Europe. *Modern Schoolman, 72*, 37–65.
Hansson, S. O. (2007a). What is technological science? *Studies in History and Philosophy of Science, 38*, 523–527.
Hansson, S. O. (2007b). Values in pure and applied science. *Foundations of Science, 12*, 257–268.
Hansson, S. O. (2013a). What is technological knowledge? In I. -B. Skogh & M. J. de Vries (Eds.), *Technology teachers as researchers* (pp. 17–31). Rotterdam: Sense Publishers.
Hansson, S. O. (2013b). Defining pseudoscience – and science. In M. Pigliucci & M. Boudry (Eds.), *The philosophy of pseudoscience* (pp. 61–77). Chicago: Chicago University Press.
Hoppe, B. (2011). The Latin artes and the origin of modern arts. In M. Burguete & L. Lam (Eds.), *Arts: A science matter* (pp. 2: 35–68). Singapore: World Scientific.
James, J. (1995). *The music of the spheres: Music, sciences, and the natural order of the universe*. London: Abacus.
Kibre, P. (1984). Arts and medicine in the universities of the later middle ages. In P. Kibre (Ed.), *Studies in medieval science: Alchemy, astrology, mathematics and medicine* (pp. 213–227). London: Hambledon Press.
Kristeller, P. O. (1980). *Renaissance thought and the arts. Collected essays*. Princeton: Princeton University Press.
Layton, E. (1976). American ideologies of science and engineering. *Technology and Culture, 17*, 688–701.
Lelas, S. (1993). Science as technology. *British Journal for the Philosophy of Science, 44*(3), 423–442.
Mayr, O. (1976). The science-technology relationship as a historiographic problem. *Technology and Culture, 17*, 663–673.
Mertens, J. (2002). Technology as the science of the industrial arts: Louis-Sébastien Lenormand (1757–1837) and the popularization of technology. *History and Technology, 18*, 203–231.
Mikkeli, H. (1997). The foundation of an autonomous natural philosophy: Zabarella on the classification of arts and sciences. In D. A. Di Liscia, E. Kessler, & C. Methuen (Eds.), *Method and order in Renaissance philosophy of nature. The Aristotle commentary tradition* (pp. 211–228). Aldershot: Ashgate.
Norström, P. (2011). Technological know-how from rules of thumb. *Techné: Research in Philosophy and Technology, 15*, 96–109.
Ovitt, G., Jr. (1983). The status of the mechanical arts in medieval classifications of learning. *Viator, 14*, 89–105.
Peckham, S. F. (1898). The genesis of Bitumens, as related to chemical geology. *Proceedings of the American Philosophical Society, 37*, 108–139.

Salmon, M. H. (2003). The rise of social anthropology. In T. Baldwin (Ed.), *The Cambridge history of philosophy 1870–1945* (pp. 679–684). Cambridge: Cambridge University Press.
Sebestik, J. (1983). The rise of the technological science. *History and Technology, 1*, 25–43.
Steele, R. B. (1900). The Greek in Cicero's epistles. *American Journal of Philology, 21*, 387–410.
Tatarkiewicz, W. (1963). Classification of arts in antiquity. *Journal of the History of Ideas, 24*, 231–240.
Tonelli, G. (1975). The problem of the classification of the sciences in Kant's time. *Rivista critica di storia della filosofia, 30*, 243–294.
Tulley, R. J. (2008). Is there techne in my logos? On the origins and evolution of the ideographic term – technology. *International Journal of Technology, Knowledge and Society, 4*, 93–104.
Van Den Hoven, B. (1996) *Work in ancient and medieval thought: Ancient philosophers, medieval monks and theologians and their concept of work, occupations and technology* (Dutch monographs on ancient history and archaeology, Vol. 14). Amsterdam: Gieben.
Weisheipl, J. A. (1965). Classification of the sciences in medieval thought. *Mediaeval Studies, 27*, 54–90.
Whitney, E. (1990). Paradise restored. The mechanical arts from antiquity through the thirteenth century. *Transactions of the American Philosophical Society, 80*(1), 1–169.
Zilsel, E. (1942). The sociological roots of science. *American Journal of Sociology, 47*, 544–562.
Zilsel, E. (2000). *The social origins of modern science* (Boston studies in the philosophy of science, Vol. 200, D. Raven, W. Krohn, & R. S. Cohen, Eds.). Dordrecht: Kluwer Academic.

Part II
The Technological Origins of Science

Chapter 3
Technological Thinking in Science

David F. Channell

Abstract Technological thinking has played a role in science throughout history. During the ancient period mechanical devices served as ways to investigate mathematical and scientific ideas. In the medieval period the mechanical clock provided scientists with a new way to conceive of time. By the period of the Scientific Revolution the clock came to play an important role in the development of the mechanical philosophy and new devices, like the air pump served as the basis for the experimental philosophy. During that period technology also came to provide a new ideology for science. In the eighteenth and nineteenth centuries the emergence of the engineering sciences played an important role in scientific thinking with thermodynamics serving as a new way to understand all scientific processes. Engineering concepts such as strain, elasticity, and vortex motion provided a way to think about electromagnetism and theories of the aether. The scientification of technology during the second half of the nineteenth century led to science-based industries which in turn led to industry-based science emerging from the industrial research laboratories. By the twentieth century the military-industrial-academic complex and the emergence of big science combined to create technoscience in which the distinctions between science and technology became blurred. The role of technological thinking in science culminated in the computer replacing the heat engine, and the clock before that, as a new model to understand scientific phenomena.

1 Introduction

In recent years there has been much debate concerning the relationship between science and technology. One of the oldest, and probably most common, ideas is the assumption that technology is dependent upon science. Since at least the second half of the nineteenth century there has been the widespread belief, particularly among scientist and the public at large, that technology simply is applied science.

D.F. Channell (✉)
School of Arts and Humanities, The University of Texas at Dallas, Richardson,
Texas 75080, USA
e-mail: channell@utdallas.edu

According to this view technology can be completely subsumed under science. Science rationalizes empirical practices that arise in older technologies and then becomes the essential source of knowledge for all modern developments in technology. Science is seen as a precondition for modern technology. On the other hand, technology simply applies scientific theories and methodologies to practical problems without contributing to, or transforming in any way, that scientific knowledge. By 1933 this model of the relationship between science and technology had become so widely accepted that visitors to the World's Fair in Chicago entered the midway under a motto proclaiming "Science Finds, Industry Applies, Man Conforms." A little more than a decade later, Vannevar Bush, who led America's research effort during World War II, set out what became known as the linear model when he said "basic research leads to new knowledge. It provides scientific capital. It creates the fund from which the practical applications of knowledge must be drawn. They are founded on new principles and new conceptions, which in turn are painstakingly developed by research in the purest realms of science" (Bush 1945, p 2).

Recently Paul Forman has argued that the primacy of science over technology is a basic hallmark of modernity but he goes on to argue that historians of technology have mostly rejected the idea that technology is simply applied science and instead have supported the view that technology is essentially independent of science (Forman 2007; Staudenmaier 1985). But by ignoring that any relationship exists between science and technology historians of technology have not appreciated the fact that recently, in an era that he labels postmodernity, the relationship between science and technology has actually reversed and there now exists a primacy of technology over science.

At the same time historians of science have often ignored or placed limits on the role of technology in science. Often times science is distinguished from technology based on a set of quadruple dichotomies – social, intellectual, teleological and educational (Hall 1959, p 4). Scientists were cerebral, or conceptual, seeking mainly understanding, not earning their income from their work, and university trained. On the other hand technologists were practical, or operational, seeking practical success, earning their wages from their trade and educated through an apprenticeship system. Based on these dichotomies very little interaction can take place between science and technology and when it does it is often seen to be initiated by the scientist and is many times restricted to the introduction of new techniques and instruments. Therefore, technology's influences on science focuses on the purely empirical which lead to new methodologies but have limited impact on the conceptual or ontological aspects of science.

While historians of technology have tended to ignore the possibility that technology could play a significant role in science and that modern science might even be characterized as applied technology, philosophers of technology have been more open to considering the role of technological thinking in science. Much of this can be traced to the influential role of Martin Heidegger's ideas in the philosophy of technology. He put forward the idea that technology as a way of thinking reveals the world to be standing-reserve (Heidegger 1977, p 17). That is, technology comes to see nature as some type of potential energy that can be "extracted and stored"

(Heidegger 1977, p 14; Ihde 2010, p 34). Because technology reveals nature to be standing-reserve, technology becomes the source of science – that is, technology is ontologically prior to science (Ihde 2010, p 37). But for Heidegger, this ontological precedent of technology over science seems to apply only when technology is seen as a way of revealing. When he turns to actual modern technology he argues that science is chronologically prior to technology. He tries to resolve this apparent contradiction by noting that the science that chronologically precedes modern technology already has within it the essence of technology and therefore technology is still ontologically prior to science (Heidegger 1977, p 22). Recently Don Ihde has argued that Heidegger had a limited understanding of the history of technology and when we begin to turn to the scholarly study of the history of technology we will find examples of technology being both historically and ontologically prior to science (Ihde 2010). As such technology can often be seen to provide more than simply a methodological impact on science but can also have a fundamental conceptual impact.

2 Ancient and Medieval Periods

2.1 Ancient Science and Technology

Throughout much of history, technology can be seen as being chronologically prior to science. Stone, copper, bronze and iron age tools all arose before any body of knowledge existed that can be called pure science, and in some cases technologies provided a conceptual basis for the establishment of science. Edmund Husserl in an appendix to his *Crisis in European Science and Transcendental Phenomenology* makes an argument that geometry and a geometrical approach to science originated in what he calls the life world (Ihde 1990, p 28). That is, through practical activities such as surveying and carpentry concepts such as lines, angles, planes and curves began to arise and become conceptualized into geometrical thinking. According to the Greek historian Herodotus, geometry arose from the needs of the Egyptians to establish boundaries of fields after the annual Nile flooding (Hodges 1970, p 132).

In the Hellenistic period the study of levers and other simple machines by people like Euclid and Archimedes served as a new way to think about mechanical and mathematical principles. While it is tempting to take a modern point of view and interpret Archimedes's work on the lever and other simple machines as an application of science to technology, the primary reason the Greeks had for studying such devices as the lever was to investigate mathematical principles through the use of physical examples. In fact one of Archimedes more famous works was entitled *Geometrical Solutions Derived from Mechanics* (Archimedes 1909). In it he said: "I have thought it well to analyze and lay down for you in this same book a peculiar method by means of which it will be possible for you to derive instruction as to how certain mathematical questions may be investigated by means of mechanics"(Archimedes 1909, p 3) As such it may be more correct to see Euclid's and Archimedes's work on the balance and the lever as an application of technology to science in the sense that simple machines were providing ways to gain a deeper knowledge of scientific principles. Similar questions have been raised about the work of Hero of Alexandria (Drachmann 1948, 1962; Hodges 1970). In his most famous book, the *Pneumatics*, he presents a large number of complex machines and statues that were powered by either air or water (Hall 1973; Landels 1978, pp 201–203). He paid particular attention to the Aristotelian idea of the impossibility of a vacuum. While the *Pneumatics* looks to be an example of the application of science to technology, several scholars have raised questions whether the described devices were ever actually constructed, or even intended to be constructed. Since none of the described devices could be considered as truly practical, it raises the possibility that they were intended to be physical examples designed to provide a concrete way to think about some scientific principle such as nature abhors a vacuum. As such, Hero's work is again more an example of technology influencing science than the other way around.

2.2 Medieval Science and Technology – The Mechanical Clock

One of the most significant technological developments that influenced scientific thinking was the invention of the mechanical clock. Although there is continuing debate concerning the origins of the mechanical clock, by the first half of the fourteenth century there were several reports of mechanical clocks in both England and Italy (Landes 1983, pp 53–58). Such clocks, driven by a falling weight, were improvements on sundials and water clocks and may have arisen in response to an interest in better time-keeping in both monasteries and the cities of medieval Europe (Landes 1983, pp 58–66). Although the new mechanical clock was more accurate than sundials and water clocks, it would not be until the late seventeenth and early eighteenth centuries that clocks, such as Christiaan Huygen's pendulum clock or John Harrison's marine chronometer, were accurate enough to be of scientific use in areas like astronomy (Landes 1983, pp 29–30, 146–152). But, before its use as a scientific instrument, the mechanical clock would influence scientific thinking by providing a new framework in which to conceive of time.

By the fifteenth century, mechanical clocks began to spread throughout Europe and began to change the way scientists conceived of time (Landes 1983, pp 77–78; Cipolla 1978, p 169). Before the mechanical clock, time was determined by physical events, usually the rising and setting of the sun. In ancient times the day was usually divided into 12 daytime hours and 12 nighttime hours. Because the amount of daylight and night change throughout the year, this meant that the length of an hour changed depending on whether it was day or night or with the season of the year (Mumford 1963, pp 15–16). Such a system was particularly suitable to an agricultural society in which work usually began with sunrise and ended at sunset. Sundials naturally marked the varying units of daylight throughout the year and water clocks could have different scales for measuring time during different seasons. But the mechanical clock beat at a regular pace independent of day or night or the season of the year. While in theory the mechanism of a mechanical clock could have been adjusted to mark off varying hours, in practice it would have been difficult to achieve. As such the mechanical clock provided a new and abstract way of measuring time (Landes 1983, pp 76–77). Mechanical time was uniform and independent of any event. Instead of time being defined as existing in some event, such as the rising and setting of the sun, these events now were seen as taking place against the backdrop of a mechanical and uniform concept of time. Without this new idea of time, much of the development of modern science would have been impossible (Cardwell 1995, pp 42–43). An experimental tradition of science would have been very difficult, if not impossible, if the measurements taken could vary depending upon the time of day or the month of the year that they were conducted.

3 The Scientific Revolution

3.1 *The Mechanical Philosophy*

By the period of the Scientific Revolution in the sixteenth and seventeenth centuries, the mechanical clock had become more than simply a tool for uniform measurements; it had become a model for understanding the natural world. During the Scientific Revolution a new philosophy of nature, that can be called the mechanical philosophy, replaced the Aristotelian philosophy that had dominated much of the Middle Ages (Mayr 1986; Boas 1952). Combining ideas drawn from Renaissance neoplatonism and the revival of Greek atomism, natural philosophers, such as René Descartes, Pierre Gassendi, Thomas Hobbes, Robert Boyle, and G.W. Leibniz, among others, formulated a variety of new scientific theories that rejected the substantial forms, the essences, and the occult forces that were central to medieval scholastic and Aristotelian natural philosophy (Mayr 1986, p 56).

The work of Descartes provided an important foundation for the mechanical philosophy (Channell 1991, pp 16–18). His idea of *cogito ergo sum* led him to make a radical distinction between mind and body. Although it was possible for him to doubt that he had a body, he could not doubt that he existed since he was doing the doubting. On the other hand if he stopped thinking there was no way to prove that he existed even though his body might still exist. This led Descartes to a rigid dualism between mind and body, or matter and spirit (Channell 1991, p 30). According to this dualism, the physical world, everything external to the human mind, was absolutely distinct from spirit or soul. This led to a new view of the natural world which was now seen to be devoid of any spirit or soul. Gravity could no longer be explained as an object's desire to return to its "natural place," and the action of magnets could not be attributed to a "magnetic soul" (Channell 1991, p 17). All matter, both organic and inorganic was entirely passive. Since totally passive matter could only interact with other matter through contact, like the gears of a machine, this new view of matter led Descartes to view the world as functioning like an automata, which for him included clocks (Mayr 1986, pp 62–63).

Mechanical philosophers such as Gassendi and Hobbes began to explain physical phenomenon in the machine-like terms of matter and motion (Channell 1991, pp 19–21). For Gassendi the motion of atoms would bring them together into corpuscles of varying shapes and sizes. If the corpuscles were smooth the material would exhibit fluid-like properties. Hot bodies would have light atoms that were easy to move while cold bodies contained heavy atoms that were difficult to move (Boas 1952, p 430). Hobbes explained the hardness and elasticity of matter by assuming the particles of a body moved in circles with the hardest bodies have particles moving fastest and in the smallest circles. Bending a body like a bow distorted the circular motion of the particles but since the particles retained their motions the shape of the bow would be restored (Hobbes 1839, pp 32–35).

The idea that the world could be best understood as functioning mechanically like a machine quickly spread during the Scientific Revolution and came to provide a model for not just the physical world but also the biological world (Channell 1991, pp 30–40). Even before Descartes put forward his mechanical philosophy, William Harvey in his *De motu cordis et sanguinis* argued that the blood circulated through the body in a single system of veins and arteries and that in order to do so the heart had to give the blood a mechanical impulse (Pagel 1967, p 52; Basalla 1962). Harvey began to speak of the circulatory system as if it were a hydraulic system with the heart acting as a pump and the veins and arteries serving as pipes (Webster 1965). Shortly after, Descartes in his *Discourse on the Method* put forward the idea that all bodily functions could be understood in terms of a machine. Having observed mechanical and hydraulic automata in the royal gardens at Saint-Germaine-en-Laye, he argued in his *Treatise of Man* that parts of the blood enter the brain and serve as a source of fluid which enters the nerves, which Descartes also saw as pipes, and the nerves carried that fluid to the muscles where the increased volume would cause the muscles to move (Rosenfield 1968, p 6). Building on the ideas of Harvey and Descartes a number of seventeenth century scientists developed iatromechanical theories. Giovanni Borelli, who studied with a student of Galileo's and Descartes's tried to explain muscular motion in terms of mechanics by treating the bones in the arms and legs as levers (Ademann 1966, p 150). Marcello Malpighi, one of Borelli's students, extended Borelli's iatromechanical research to other parts of the body. He discovered that the tongue contained pores leading to sense receptors which confirmed Galileo's theory that taste arose from particles penetrating the upper part of the tongue and that different tastes arose from the different mechanical shapes or motions of the particles of food (Belloni 1975, p 97). Malpighi also suggested that the glands were essentially sieves whose purpose it was to separate particles from the blood and send them to a secretory duct.

The fact that the mechanical philosophy could be used to explain both the physical world and the organic world led to the idea that the entire universe was a gigantic clockwork and God was the clockmaker. A leading exponent of the idea of a clockwork universe was the seventeenth century British chemist Robert Boyle who argued that the universe was "like a rare clock, such as may be that at Strasburg, where all things are so skillfully contrived, that the engine being once set a-moving, all proceed according to the artificer's first design, and the motions ... do not require the peculiar interposing of the artificer ... but perform their functions upon particular occasions, by virtue of the general and primitive contrivance of the whole engine" (Channell 1991, p 22) The great Strasbourg clock became an appropriate symbol of the clockwork universe since it not only contained a large dial depicting the motions of all of the planets, but also a series of 12 jackworks representing human figures in addition to a mechanical rooster.

The Scientific Revolution culminated with a debate that focused on the idea of a clockwork universe. Isaac Newton had raised questions as to whether the universe could be considered a perfect clockwork since he believed that the gravitational effects of comets would interfere with the orbits of the planets and require God's intervention (Mayr 1986, pp 97–120). In a series of letters between Leibniz and the Reverend Samuel Clarke, a student of Newton's, Leibniz argued that Newton's view of the universe implied God was in imperfect clockmaker who needed to continually wind, clean and mend the clockwork universe (Koyré and Cohen 1962; Shapin 1981). While the Clarke-Leibniz debate was not resolved in their lifetimes, by the end of the eighteenth century Pierre Simon Laplace, the French mathematician, was able to show that the Newtonian universe was self-correcting. The idea that the universe functioned like a machine implied that the study of nature was essentially a study of technology (Channell 1991, pp 26–27).

3.2 The Experimental Philosophy

During the Scientific Revolution technological thinking played a role in science not only in the formulation of the mechanical philosophy but in the establishment of an experimental philosophy (McMullin 1990). Before the Scientific Revolution the most significant approach to obtaining knowledge about the natural world was through a deductivist or mathematical approach which began with a series of self-evident principles, propositions, or axioms, and then proceeded to some conclusions through the application of some formal set of rules, logic, or procedures. Ancient and medieval philosophers placed little value in what we would call the experimental method. Aristotelian philosophy established a rather rigid distinction between and natural and the artificial. For Aristotle, in order to understand the essence of some natural phenomenon one had to study that phenomenon in its natural state. Also an experimental approach was somewhat limited in the ancient and medieval worlds because of the belief that the universe was divided into two fundamentally different regions – the terrestrial and the celestial – each of which had its own set of physical laws. In such a system no experimental situation could be created in the terrestrial region that would provide any information on the functioning of the celestial region. The social and intellectual changes that were taking place during the Scientific Revolution led to a new attitude toward experimentation. Renaissance neoplatonism undermined Aristotelian philosophy and the Copernican system of astronomy erased distinctions between the terrestrial and the celestial and led to a unified view of the universe.

During the seventeenth century a new experimental philosophy was encouraged by the writings of the English statesman and philosopher, Francis Bacon (Zagorin 1998; Briggs 1989; Martin 1992). In three significant works, *The Advancement of Learning*, *The Great Instauration*, and *The New Organon*, he put forward the idea that an understanding of natural philosophy had to begin with empirical data, then proceed through a process of induction to an understanding of the material and efficient causes, and culminated in the discovery of laws of nature (Zagorin 1998, p 64). Bacon has often been portrayed as advocating a simple collection of facts but his method depended much more upon an experimental approach to nature rather than simple observations. He said the important data about nature "cannot appear so fully in the liberty of nature as in the trials and vexations of art" (Zagorin 1998, p 62). That is, like the lawyer that he was, Bacon believed that nature did not give up her secrets without some form of interrogation. This differed greatly from Aristotle's view that natural phenomena had to be in their natural state. Bacon's view opened the possibility of a study of nature in an artificial state that had been created through technology. Some natural philosophers such as Descartes and Hobbes argued strongly against the experimental philosophy, but others such as Galileo, Boyle and Newton incorporated experiments into their idea of a mechanical philosophy.

Experimental technologies influenced scientific thinking in a number of ways during the seventeenth century. Instruments could extend the senses, they could serve as a mediator between humans and the world, they could help to reason about the world, and they could confirm or demonstrate a theory (Hankins and Silverman 1995, pp 1–13). But all of these roles for instruments became the subject of much debate. The use of the telescope by Galileo helped to confirm the Copernican system of astronomy by extending human sight so as to be able to see the moons of Jupiter and the mountains on the moon. The use of the microscope by Robert Hooke and Malpighi led to the discovery of cells in plants. But these discoveries were often not accepted at face value. Although he was one of the founders of the experimental philosophy, Bacon believed that instruments like the telescope distorted the senses and therefore might not be a source for scientific knowledge. On the other hand Hooke believed that instruments helped to overcome the "infirmities" of the senses, but this raised the issue of the fallibility of the senses (Shapin and Schaffer 1985, pp 36–37). Such debates over the role of technological instruments in science reflected the fact that in order for experiments to be accepted as revealing some truths about nature, new standards had to be established concerning what constituted a scientific experiment, how an experiment was to be performed, what was the relationship between experiments and theories, and how was an experiment seen as a legitimate way to understand the natural world (Shapin and Schaffer 1985, p 18).

As Steven Shapin and Simon Schaffer have shown in their book *Leviathan and the Air Pump*, one of the most important debates during the Scientific Revolution concerning the role of a technological device in scientific thinking took place over the air pump (Shapin and Schaffer 1985). As we have seen, Robert Boyle was a major supporter of the mechanical philosophy and one of his most important contributions was to show through experiment that phenomena could be explained by some mechanical hypothesis. The air pump played a role in the debate over an experiment conducted by Evangelista Torricelli who filled a glass tube, sealed at one end, with mercury and inverted the tube into a glass dish. While some of the mercury flowed out, a column of mercury remained in the tube. Scientists debated the "Torricelli space" that existed above the mercury in the glass tube and how that space was formed. Some argued that the weight of the air acting on the mercury in the dish kept all of the mercury from flowing from the tube and the resulting space was a vacuum. Others, denying the existence of a vacuum, argued that the mercury was held up in the tube because the air in the space above it had reached the limit of its expansion (Shapin and Schaffer 1985, p 41).

Boyle, using a newly invented air pump, was able to create an artificial void in which he could conduct a variety of experiments. In one experiment he placed Torricelli's device into the chamber of his air pump and noted that with each stroke of the air pump the height of the column of mercury decreased until it almost reached the level of the mercury in the dish that held the tube (Shapin and Schaffer 1985, pp 41–44). Based on his experiment Boyle concluded that the column of mercury could not be supported simply by the weight of the air since its weight was only a few ounces. Instead, Boyle postulated that air had a spring-like quality, or pressure, that resisted being compressed and air expanded when it was not contained by some force. It was the normal pressure of the air that supported the column of mercury in the Torricelli device and when the quantity of air decreased, after being pumped out, the pressure or expansive quality of the air was reduced and the column of mercury fell. Boyle went on to conduct more that 40 different experiments using the air pump, including experiments on cohesion, combustion and animal physiology.

From the modern point of view Boyle's experiments and the conclusions drawn from them seem straightforward but Shapin and Schaffer have shown that the interpretation and acceptance of such experiments was anything but straightforward and much of the debate focused on the role of a mechanical device in scientific thinking (Shapin and Schaffer 1985, pp 60–65). At the time of his experiments very few air pumps existed and most had significant leaks, which made it difficult for others to reproduce Boyle's experiments or interpret exactly what took place during the experiment. Philosopher Thomas Hobbes argued that while experiments could provide a number of particular facts about nature, he did not see how such particular facts could lead to knowledge concerning causes (Shapin 1996, pp 110–111).

In order for scientists to accept the role of a technological device in scientific thinking, new technical, social and rhetorical practices had to be established for dealing with experiments. First, the air pump had to be recognized as an impersonal device that produced data or facts that were independent of the human observers. Good or bad data became attributed more to the machine than to those conducting the experiments. Second, Boyle conducted many of his experiments in public, making the process a social enterprise in which those witnessing the experiments played a role in their validation. Shapin suggests that gentlemanly codes of honor, honesty and truthfulness played an important role in the acceptance of such experimental results (Shapin 1996, p 88). Finally, Boyle not only conducted his experiments in public but he also created written reports so that individuals not able to witness the experiment directly could have enough detailed information that they could be considered "virtual witnesses" (Shapin 1996, p 108; Shapin and Schaffer 1985, pp 60–65). The creation of narratives that allowed such virtual witnessing required science to adopt new rhetorical techniques that would allow readers to feel that they had in fact observed the experiment.

3.3 A Technological Ideology in Science

Possibly one of the most important changes brought about by the role of technological thinking in science during the Scientific Revolution was the development of a new ideology of science. Throughout the ancient and medieval periods science was seen as a branch of philosophy and its primary goal was to obtain knowledge or truth about nature simply for its own sake. With a few exceptions, science was not seen a practical or useful. Science was often seen as both socially and intellectually distinct from technology although some barriers between the two areas began to be overcome during the Middle Ages. But during the early modern period the social, political, economic, and intellectual forces that gave rise to the Scientific Revolution started to reshape the ideology of science so that it began to more closely resemble the goals and values that we associate with technology. Scientific knowledge was no longer valued simply for its own sake, but science began to be seen as useful, practical and powerful. Often studies of the Scientific Revolution ignore the fact that it emerged out of a period of intense technological activity that produced ocean-going ships, guns and gunpowder, and movable type printing. As we have seen, technology played a significant role in the mechanical and experimental philosophies that were at the core of the Scientific Revolution.

The person most responsible for making explicit a new ideology of science that was modeled on technology was Francis Bacon. Although Bacon is often better known for his contributions to the development of a new methodology of science, his most significant contribution to modern science may have been his formulation of a new ideology of science. Like many others of his era, Bacon believed that the millennium was at hand and that science could help to restore the human condition to one which existed before the Fall (Noble 1997, p 50). Throughout his earlier works Bacon argued that the goal of humans was to both know and master nature (Zagorin 1998, p 78). But Bacon's most significant contribution to a new ideology of science was contained in his work *The New Atlantis* (Bacon 1937). Organized as a utopian work, the book was a fable that described a group of European travelers who were driven off course and shipwrecked on a remote island that held a Christian utopian community (Zagorin 1998, p 171). At the center of the community was an institution named Salomon's House, or the College of the Six Days of Work. As the visitors would discover, Salomon's House was not a political institution but rather a great research institution with laboratories, equipment and instruments to conduct experiments and make inventions (Bacon 1937, pp 480–481). But the purpose of Salomon's House was not simply to conduct pure research in order to gain knowledge about the natural world; rather its goal was to conduct group research in order to bring about improvements in society. Work was conducted on mining, fishing, raising livestock, growing fruits and vegetables, metallurgy, medicine, brewing and baking, weaving and dyeing textiles, optics and acoustics, new sources of power, ship design and navigation, and flight (Bacon 1937, pp 481–488). Through his writings, especially *The New Atlantis,* Bacon put forward a new ideology of science, one that was no longer a solitary, contemplative search of truth but one that was collaborative, practical and a source of power. This new ideology of science was much closer to technology and would reshape science. The idea that scientific research could be a collaborative effort like technology was a change in the way scientific research had be done in the past and would serve as a model for the Royal Society of London and later for the modern-day industrial research laboratories. Also there was now the idea that science should turn to technology for its research agenda. For example, Galileo begins his *Discourse on the Two New Sciences* with the statement: "The constant activity which you Venetians display in your famous arsenal suggests to the studious mind a larger field for investigation, especially that part of the work which involves mechanics" (Galilei 1954, p 1).

4 The Eighteenth and Nineteenth Centuries

4.1 The Engineering Sciences

During the eighteenth and nineteenth centuries, technological thinking had a significant impact on science through the emergence of the engineering sciences (Channell 2009; Channell 1989, pp xvi–xxiii). The creation of the engineering sciences was a response to the technological developments that had arisen during the Industrial Revolution. With the invention of steam engines, railways, ocean-going iron-hulled ships, and large scale iron bridges, it became impractical and uneconomical for engineers to use traditional rule-of-thumb or trial-and-error techniques, so they began to turn to science for assistance. But, much of the science that emerged from the Scientific Revolution was of little direct use in technology. Newtonian mechanics might explain the forces acting between two point atoms but it did not help in determining how an iron beam might act under a complex load in a bridge. Boyle's law explained the relationship between pressure and volume in an ideal gas, but it was of little use in describing how steam acted in a working steam engine. The Bernoulli equation of classical fluid mechanics had limited application in describing real fluids undergoing non-laminar flow. In its first editorial the British journal, *The Engineer*, recognized the existence of a new type of science when it said: "There is a science of the application of science, and one of no minor importance. The principles of physics ... would remain only beautiful theories for closet exercise but for the science of application" (Healey 1856, p 3). While the engineering sciences would at first be seen as the application of science to technology, they would also come to play an important role in the application of technological ideas to science. Among the fields that began to develop into engineering sciences were the strength of materials, the theory of elasticity, the theory of structures, the theory of machines, hydrodynamics, fluid mechanics and thermodynamics. As intermediate and independent bodies of knowledge existing between science and technology, the engineering sciences served as a form of translation between the two areas. In doing so the engineering sciences developed a series of concepts that were related both to the ideal concepts of pure science and the more practical concepts used to describe technological artifacts. Concepts such as stress, strain, efficiency, work, entropy, and streamlines, among others, were neither purely scientific nor purely technological concepts but a combination of the two.

While the development of the engineering sciences had the greatest impact on technology, turning it into more of a science, the engineering sciences also played a significant role in scientific thinking. In particular, the conceptual framework of the engineering sciences was often used in the development of late nineteenth century scientific theories dealing with energy, electromagnetic fields, the aether, atoms, and the electron.

4.2 Thermodynamics

One of the most significant roles played by an engineering science in scientific thinking arose out of studies of the steam engine which led to the development of thermodynamics and then the broad science of energy (Cardwell 1971; Hills 1989). The first practical steam engine had been invented by Thomas Newcomen of Cornwall in 1712 with little input from science. As has often been said, science owed more to the steam engine than the steam engine owed to science. With demands for new sources of power brought about by the Industrial Revolution, there was an increased interest in improving the engine and gaining a better understanding of the scientific principles that lay behind the steam engine.

In 1769 James Watt patented a group of improvements, such as a separate condenser, that made the steam engine much more efficient. In order to better understand his engine Watt conducted a series of experiments. His use of the expansive power of steam along with condensing steam made it difficult to calculate the power of a given engine without knowing how the pressure of steam was dropping inside the cylinder. In 1796 John Southern, one of Watt's assistants, developed a simple device to measure the pressure inside the cylinder throughout its cycle by using a moving piece of paper attached to a marker (Cardwell 1971, pp 79–81). The resulting "indicator diagram," which was essentially a pressure-volume (P-V) curve, allowed Watt to calculate the power of his engines. By the middle of the nineteenth century the P-V diagram would become a fundamental element of the science of thermodynamics.

The study of the steam engine was further stimulated by the development of new types of engines after Watt's patent ended in the early nineteenth century. Studies of the new engines seemed to confirm that the most efficient engines were those invented by Arthur Woolfe that were high pressure expansive engines and also used condensers, but there was no theory to explain why this was so (Cardwell 1972, p 90). When the new engines were introduced into France, Sadi Carnot became interested in explaining the engine's superior efficiency and in doing so he went beyond the specific problem of explaining the Woolfe engine and developed a general theory of heat engines regardless of the type of engine or the working substance. He put forward his theory in 1824 in his book *Réflexions sur la puissance motrice de feu*. Many have seen this book as the beginning of what William Thomson (later Lord Kelvin) would label in 1849 as the new science of thermodynamics (Smith 1998). Although Carnot's theory of an ideal engine, or Carnot cycle, is still accepted today, it was based on the now rejected caloric theory of heat which saw heat as a substance. The caloric theory allowed Carnot to analyze heat engines as analogous to already well understood theories that applied to water power. Using the analogy, he put forward the principle that heat should flow (or fall) from the highest to the lowest possible temperature and the principle that there should be no useless flow of heat (Cardwell 1971, p 193; Cardwell 1972, p 93). With these two principles Carnot formulated the concept of an ideal engine which followed what became known as a Carnot cycle (Cardwell 1971, pp 194–195). Because his cycle was reversible his ideal engine represented the most perfect engine that could be conceived since if any more efficient engine existed it could be used to drive a Carnot engine in reverse and use the heat to run that more efficient engine resulting in perpetual motion (Cardwell 1972, p 94). Therefore the Carnot cycle became an ideal standard against which all real engines could be compared. Carnot's theory had little impact on scientific thinking until 1834 when Emile Clapeyron reformulated Carnot's theory into a mathematical format in which the cycle was presented in terms of a P-V, or indicator diagram (Cardwell 1971, pp 220–221; Smith 1998, pp 44–45). But the full impact of Carnot's theory on the development of the science of thermodynamics did not come until the idea emerged of the interconvertability of heat and work which contributed to the concept of the conservation of energy (Cardwell 1972, p 96).

The recognition that heat was not always conserved but could be converted into work was itself shaped by technological thinking. Count Rumford (Benjamin Thompson) had raised doubts about the caloric theory of heat when he observed that the boring of cannons seemed to be able to produce and unlimited amount of heat and that the duller the tool the more heat was released. This was the opposite of what was to be expected if caloric was being released from the metal shavings produced by a sharp tool. By the 1840s Julius Mayer and James Prescott Joule firmly established the mechanical theory of heat. Joule's thinking in particular was influenced by technological ideas (Smith 1998, pp 53–73; Cardwell 1971, pp 231–238; Cardwell 1976, pp 674–686). He was especially interested in the possibility that electrical motors might be more efficient than the steam engine. His studies of electrical motors led Joule to become interested in the relationship between heat and work. He had noted that the resistance to the flow of electricity in a wire produced heat. In a circuit powered by a battery it seemed clear at the time that the heat arose from chemical activity taking place in the battery, but heat was also generated when the circuit was powered by a magneto (a device in which a coil of wire and a permanent magnet are rotated relative to one another thus producing electricity through electromagnetic induction). Heat produced by a battery could be explained by the idea that caloric was already in the battery and was simply released through chemical activity, but there was only mechanical activity taking place in the magneto. This led Joule to conclude that heat was being produced by mechanical activity and that heat was a mechanical phenomenon rather than a substance. After a series of other experiments, Joule established that there was a fixed mechanical equivalent of heat and that heat and work could be mutually transformed into one another.

In order to establish a new science of thermodynamics Joule's idea of the mechanical equivalent of heat had to be reconciled with the Carnot-Clapeyron theory of the ideal heat engine based on the older caloric theory of heat which in turn was based on the concept of the conservation of heat. The solution to this problem arose with the development of the idea of the conservation of energy. While Thomas Kuhn has argued that at least 12 scientists and engineers contributed to the idea of the conservation of energy, the formulation of a science of energy as a new unifying framework for science and engineering was primarily the result of the work of a group of Scottish scientists and engineers, including William Thomson (later Lord Kelvin), W.J.M. Rankine, James Clerk Maxwell, and Peter Gutherie Tait, along with the German scientist and engineer, Rudolf Clausius (Kuhn 1969; Smith 1998). In 1850 Clausius published a paper in which a put forward the idea that the theories of Carnot and Joule could be reconciled (Harman 1982, pp 52–55; Smith 1998, pp 97–99; Cardwell 1971, pp 244–249). He argued that the fundamental element of Carnot's theory was that during a cyclical process work was done when heat passed from a hotter body to a cooler body and this principle could be accepted even if heat was not conserved in the process (Smith 1998, p 97). Clausius made two assumptions – that heat and work were equivalent (Joule's principle) and that during a cyclical process some heat was converted into work while another portion was simply transmitted from a higher temperature to a lower one (a revision of Carnot's principle). These two principles would later form the basis for the first and second laws of thermodynamics.

Shortly after the publication of Clausius's paper, Thomson, in discussions with Rankine, began to argue that there were some problems with Clausius's revision of Carnot's principle (Smith 1998, p 106; Cardwell 1971, pp 254–257). In a series of papers published between 1851 and 1855 Thomson focused on the problem of irreversibility – that is, dealing with the heat of conduction that simply passed from a higher temperature to a lower temperature but did not produce any mechanical effect (Harman 1982, p 56). Thomson explained this problem in terms of a dynamical theory of heat – that is, heat was the motion of particles of matter. If this was the case, then heat that was conducted from a hot body to a cooler body was not lost, rather such heat simply dissipated through the cooler body causing its particles to increase their motions but not be converted into usable work. For Thomson this dissipation of heat was just a fundamental as the mechanical transformation of heat into work. This meant that in a heat engine heat was not conserved, as Carnot believed, and it was not totally converted into work, as Joule implied. Rather some of the heat was converted into work while the rest of the heat was dissipated.

During the 1850s Thomson and Rankine began to reformulate the laws of thermodynamics in terms of the new concept of energy (Harman 1982, pp 58–59). In doing so they expanded the idea of thermodynamics from a science of heat engines to more fundamental laws of nature. Although the term energy had a long history, it had mostly been used in rather vague and imprecise ways. Both Rankine and Thomson began to argue that the term energy could be used as a basis for understanding all processes in nature including mechanics, chemistry, thermodynamics, electricity, magnetism and light. By 1853 Rankine reformulated earlier ideas on the conservation of force and Thomson's idea of the principle of mechanical effect into "the law of the conservation of energy" which stated that the total energy in the universe is unchangeable (Smith 1998, p 139). With this reformulation, the conservation of energy became the first law of thermodynamics which came to be seen as applying to all physical phenomena, not simply heat engines.

While the law of the conservation of energy became one pillar of the science of energy, the law of the dissipation of energy provided the second pillar of the new science (Cardwell 1971, pp 260–276). In 1854 Clausius began to reformulate his ideas concerning the dissipation of heat (Smith 1998, pp 166–167; Harman 1982, pp 64–66). He came to see that the transformation of heat into work could be related to the work lost through the dissipation of heat and showed the dissipated heat was equivalent to the work required to move that quantity of heat from a cooler temperature back to the original temperature. In 1865 Clausius introduced the term entropy (from the Greek word for transformation) to refer to the equivalence value of the transformation of heat (Smith 1998, pp 167–168; Cardwell 1971, p 273).

Both Rankine and Clausius showed that for reversible processes, such as a Carnot cycle, the total change in entropy would be zero, but Clausius also applied the concept to irreversible processes, such as would be encountered in actual heat engines. In such cases he concluded that the entropy would always increase. With his new formulation of entropy, Clausius was able to formulate the two laws of thermodynamics as: "the energy of the universe is a constant," and "the entropy of the universe tends to a maximum" (Smith 1998, p 168). As formulated by Clausius, the laws of thermodynamics governed all physical phenomena, not simple heat engines. During the second half of the nineteenth century thermodynamics began to be applied to a wide range of phenomena beyond heat engines. Chemists, such as Marcelin Berthelot and Josiah Willard Gibbs began applying thermodynamic principles to chemical reactions (Smith 1998, pp 260–262, 302). By the end of the century the application of thermodynamics to light and blackbody radiation would lead to the development of quantum theory (Kuhn 1978). Finally, Albert Einstein modeled the structure of his theory of special relativity on the science of thermodynamics by using just two principles – a relativity principle and a light principle – similar to the way thermodynamics was based on just two laws (Holton 1973, pp 167–169). By the second half of the nineteenth century the universal nature of the laws of thermodynamics resulted in the model of the steam engine, or heat engine, replacing the clock as a model for understanding natural phenomena (Brush 1967).

4.3 Electromagnetism

Thermodynamics was not the only engineering science to play a role in pure scientific thinking. The development by Michael Faraday of the concept of a field as a way to think of electromagnetic phenomena led to the idea that fields might be associated with the luminiferous aether that was assumed to fill all space (Cantor and Hodge 1981). William Thomson used concepts drawn from the engineering sciences to explain the relationship of electromagnetic fields to the aether. At first he suggested that fields might be associated with strains in the aether. The concepts of stress and strain had emerged in the late eighteenth and early nineteenth centuries from a new interest in the strength of materials and the theory of elasticity brought about by large scale building projects associated with the Industrial Revolution. In such projects the Newtonian idea of a force acting between points was of little use in analyzing how material beams and foundations acted in real buildings and bridges, but the concepts of stress and strain provided more useful ways to analyze material structures. Later in order to explain Faraday's discovery that a magnetic field could cause the rotation of the plane of polarization of light, Thomson suggested that all matter might simply be smoke-ring like rotations of the aether (Harman 1982, pp 82–83). The interest in vortex-like motions of a fluid had arisen from new problems in hydrodynamics associated with designing large ships and harnessing water power with waterwheels and water turbines. Both Rankine and Hermann von Helmholtz speculated that matter might be some form of vortex or smoke-ring type of motion of the aether.

The ideas of Faraday and Thomson were synthesized and brought to fruition by James Clerk Maxwell. In his early work he was influenced by Thomson and Rankine and developed a hydrodynamic model in which electric currents were seen as analogous to the motion of some type of fluid. Later in his paper "On the Physical Lines of Force," Maxwell developed an intricate mechanical-technological model of the aether that would lead him to conclude that vibrations of electromagnetic fields were similar to the propagation of light (Smith 1998, pp 223–228). Drawing on Rankine's idea of matter as composed of molecular vortices of aether and on recent developments in water turbines, Maxwell began to apply the model of vortex motion not only to matter but to the aether itself. In order to explain how vortices of aether, representing a magnetic field, could all rotate in the same direction, Maxwell drew upon engineering and suggested that between the vortex cells there were small particles rotating in the opposite direction, acting the same way that an idle wheel did in a machine (Smith 1998, pp 225–226; Harman 1982, pp 89–92). The motion of these idle wheels would be associated with electricity and provided a way to envision how electricity and magnetism were intimately connected (Wise 1990, p 351).

Maxwell's technological models led to two revolutionary conclusions. First, electricity was not a phenomenon confined to a conductor but something that could exist throughout space. If electricity was the motion of small idle wheels inside a wire, that rotary motion would eventually be transmitted to the space surrounding the wire causing layer upon layer of idle wheels outside the wire to rotate. Second, given its mechanical properties, Maxwell's model of the aether was capable of transmitting waves through it. After calculating that the speed of such waves would be similar to the speed of light Maxwell concluded that light was simply an electromagnetic vibration (Harman 1982, p 93). By 1865 Maxwell began to reformulate his theory of electricity and magnetism so that it no longer depended upon any specific mechanical model, but technology certainly played a role in his initial thinking and his visualizations of the phenomena.

Although Maxwell's theory came to be formulated in a way that was independent of any mechanical model, a number of his followers continued to use mechanical or engineering science models. In the late 1870s and early 1880s George FitzGerald developed an electromagnetic theory that pictured the aether as an elastic solid (Smith 1998, p 290; Harman 1982, pp 98–99). In later papers he argued that both matter and aether were composed of vortex motions, with matter represented by vortex rings and the aether by vortex filaments filing space (Harman 1982, p 99). As a way of understanding his model, FitzGerald proposed a mechanical analogy composed of rubber bands and wheels. By the late 1880s he proposed a hydrodynamic model of the aether that he labeled a "vortex sponge." Such a model allowed him to draw upon new developments in engineering science, such as equations of fluid dynamics, to analyze the aether. About the same time Thomson tried to combine aspects of elastic solid models of the aether and vortex fluid models. In place of his vortex model he put forward a model of the aether based on a series of gyroscopes connected to one another by a series of springs. Such a model would explain how the aether could sustain both transverse electromagnetic waves and could cause rotational effects such as the magnetic rotation of the plane of polarization of light (Harman 1982, pp 100–101).

By the 1890s Joseph Larmor combined ideas of the aether as vortex rings and the aether as an elastic solid to develop a theory of the aether as a "pure continuum," whose sole properties were elasticity and inertia (Harman 1982, p 102). Rather than being composed of matter, Larmor argued that aether was prior to matter and that matter was some structure in the aether (Harman 1982, p 102). In his *Aether and Matter* Larmor speculated that centers of strain in the aether could have properties of an electric charge and could be considered "electrons." Although Larmor's electrons differed from the modern conception and Einstein's special theory of relativity would eliminate the need of the aether, technology, especially concepts from the engineering sciences, played an important role in thinking about scientific theories during the second half of the nineteenth century.

4.4 From Science-Based Industry to Industry-Based Science

Also during the second half of the nineteenth century technology was playing an important role in thinking about experimental science. The scientification of technology during the nineteenth century led to what some have called a second industrial revolution which in turn led to the rise of science-based industries (Böhme et al. 1978). Two of the most significant science-based industries that contributed to the second industrial revolution were the chemical and electrical industries. Their rise depended upon new scientific developments such as the chemical revolution of the late eighteenth and early nineteenth centuries and the discovery of electromagnetism and electromagnetic induction during the first half of the nineteenth century. At the center of these new science-based industries was the industrial research laboratory. Beginning first in the chemical industries and then later in the electrical industries, the industrial research laboratory was created in response to the needs of continuous innovation, obtaining new patents, and protecting existing patents (Meyer-Thurow 1982, pp 367–370; Fox and Guagnini 1998–1999, p 260).

While industrial research laboratories in the chemical industries can often be seen as simply an example of applying science to technology, some important differences emerged in the industrial research laboratories that arose out of the electrical industries that not only made technology more scientific but began to make science more technological. One of the most important differences between industrial research in the chemical and electrical industries was the integration of pure and applied research in the electrical laboratories. In the nineteenth century chemical industries much of the fundamental research was still left to the universities and the industrial laboratories focused more on applied research. But because of the newness of scientific research in the field, the electrical industries required both pure and applied research in order to improve their products and processes (Reich 1985, p 240). As such, laboratory leaders were willing to support a certain level of undirected fundamental research. More importantly, within the electrical industrial research laboratories the lines between pure and applied research and the lines between scientists and engineer began to disappear. Individuals trained as scientists often did engineering work and those trained as engineers often did scientific work, and even more often scientists and engineers did both types of work.

One example of a new relationship between science and technology was the work done at the General Electric Research Laboratory (Reich 1985, p 64). Founded in 1900, the laboratory focused on problems of Edison's incandescent bulb which was a delicate object with a not very long life and was not terribly efficient. In order to compete with metal filament bulbs being developed in Europe, work at the laboratory began to focus on tungsten (Reich 1985, p 74). A group under William Coolidge, who held a Ph.D. in physics from Leipzig, conducted fundamental research on tungsten and discovered that tungsten, unlike other metals, became less brittle when it changed from a crystalline to a fibrous state (Reich 1985, pp 77–80). This led to a method to make tungsten ductile so that it could be drawn into thin filaments for light bulbs. Coolidge and his team were not simply applying scientific knowledge to a practical problem; instead they had to create the scientific knowledge needed to solve a practical problem (Reich 1985, p 120). The scientific knowledge itself was the result of a practical problem.

The new tungsten bulb was further improved by Irving Langmuir, who held a Ph.D. in physical chemistry from Göttingen. He began research into the cause of the blackening of the inside of the bulb (Reich 1985, pp 120–127). Rather than studying the problem of blackening directly, Langmuir undertook a study to investigate the basic scientific principles of incandescent lighting. He discovered that blackening was not the result of residual gas in the bulb but rather the high temperature of the filament caused tungsten to "evaporate" and become deposited on the inside of the bulb. He concluded that the introduction of a gas like nitrogen (and later argon), could inhibit the blackening. Like Coolidge, Langmuir did not simply apply science to technology but he had to create the science needed to solve a practical problem. Langmuir's work with the light bulb, a technological object, led to new scientific thinking. His studies of the problem of blackening led him to study how heat could cause electrons to be emitted from a filament – so-called thermionic emissions. These studies of the physical processes taking place inside the incandescent bulb eventually led to Langmuir being awarded the 1932 Nobel Prize in chemistry, the first industrial scientist to win such an award.

In the industrial research laboratories of the electrical industries there was another dimension to the role of technological thinking in science. Wolfgang König, in a study of the electrical industries in Germany before World War I, argues that while the electrical industries employed significant numbers of scientists with academic backgrounds, much of the academic science that pertained to the electrical industries was often generated by individuals who had previous practical experience before they entered the Technische Hochschulen (König 1996, p 87). As such, academia played a larger role in the dissemination of new knowledge than in the production of that knowledge. Given these circumstances, König argues that it is more correct to characterize what took place in the electrical industries as industry-based science (König 1996, p 73, 100).

5 The Twentieth Century

5.1 *Technoscience*

During the twentieth century technological thinking played an increasing role in science. In fact the interdependence between science and technology developed to a point where the distinctions between the two areas was beginning to disappear and in the place of individual disciplines of science and technology there emerged the new concept of a single integrated realm of knowledge that some have labeled technoscience (Latour 1987). Popularized by Bruno Latour, technoscience is often used in a variety of ways. Latour uses the term to refer to the elimination of distinctions between the notion of science as a pure, ideal, disinterested activity, and the notion of science as a practical activity shaped by societal forces (Latour 1987, pp 174–175). Some trace the idea to Heidegger's argument that if technology causes us to view the world, including nature, as a "standing reserve," then science is no longer concerned with pure knowledge but becomes an instrument that is only "fulfilled as technology"(Heidegger 1977, pp 3–35; Salomon 1973, pp xv–xvi). Somewhat following Heidegger, Jean-François Lyotard argued that: "In technoscience, technology plays the role of furnishing the proof of scientific arguments" (Sassower 1995, p 24). Raphael Sassower notes that the concept of technoscience goes beyond the traditional notion in which science is simply implemented by technology. Instead he argues that technoscience represents a new situation in which technology cannot exist without science but also where "there is no science without technology" (Sassower 1995, p 4, 24).

What is called technoscience did not come into is full existence until the second half of the twentieth century but its roots can be traced back to the first half of the century. The interdependence of science and technology that emerged from science-based industries and industry-based science during the nineteenth century would continue into the twentieth century. But during the twentieth century another factor would come to play a defining role in the emergence of technoscience – the active involvement of national governments and the role of politics in shaping scientific and technological development. Of course ever since the rise of the modern nation-states during the fifteenth and sixteenth centuries, government and politics have been active in promoting science and technology through patent systems, academies of sciences, and support of education. What distinguished the twentieth century was the breadth, scale, and explicit nature of such support. Also, throughout most of history politics has had a more direct impact on technology than on science, but during the twentieth century politics began to play a much greater and more explicit role in the development of science. Rather than science focusing on knowledge for its own sake, twentieth century science "operates as a technique like any other" and becomes "the manipulation of natural forces in the light of political decisions" (Salomon 1973, p xx). As such, technosciences' blurring of the distinctions between science and technology leads to a "new relationship between knowledge and power" (Salomon 1973, p xx).

5.2 The Military-Industrial-Academic Complex

For many scholars, the key factor in establishing a new relationship between knowledge and power that characterized technoscience was the role played by war during the twentieth century. While warfare has existed throughout history, warfare during the twentieth century was significantly different from previous wars. Primarily what had changed was the scale of warfare. Twentieth century wars were "total wars," fought world-wide in which the distinctions between civilian and military broke down. The issue of the relationship between science, technology and the military has been the subject of much scholarly debate (Roland 1985), but recent scholarship has shown that during the twentieth century the two world wars and the continuing state of war that existed during the Cold War certainly played a significant role in transforming science and technology into technoscience.

The relationship between science, technology and the military led to two important developments that helped to shape technoscience. The first development was the emergence of a military-industrial complex, a term coined by Dwight Eisenhower in his farewell address as president (Roland 2001). It referred to the establishment of a permanent private defense industry whose primary client was the military. Although Eisenhower saw the military-industrial complex as emerging in the 1950s, the roots can be traced back to the period of World War II. Also, the term military-industrial complex might be more correctly labeled the military-industrial-academic complex because it not only established a new intellectual and institutional relationship between technology and the military but it also created a new interdependence between academic science and the military, establishing what might be called a science-based military (Leslie 1993). In doing so it helped to further blur the distinctions between science and technology (Kay 2000, pp 10–11).

Because of wartime needs during both World War I and World War II, governments played an essential role in creating what would become the military-industrial-academic complex. Governments sponsored organizations such as the Kaiser Wilhelm Institute and the Physikalish-Technische-Reichsansalt in Germany, the National Physical Laboratory, Aeronautical Research Council and the MAUD Committee in Britain, and the Chemical Warfare Service, the National Advisory Committee on Aeronautics, the Naval Consulting Board, the Office of Scientific Research and Development and the Manhattan Project in America. These organizations would connect academic and industrial researchers with some of the world's leading corporations, including IG Farben, Krupp, Telefunken, BASF, British Marconi, General Electric, AT&T, Union Carbide, and Eastman Kodak. By the end of World War II the results of this military-industrial-academic complex had produced a number of winning weapons including radar, code-breaking computers, and the atomic bomb.

This new science-based military led to changes in science that made science more like technology (Capshew and Rader 1992, pp 8–9). The important military role played by the development of nuclear weapons and commercial nuclear reactors during the Cold War led to significant changes in academic science. As Stuart W. Leslie has argued, World War II and the Cold War dramatically transformed the nature of American physics through new sources of funding such as the Atomic Energy Commission, the Office of Naval Research and the National Science Foundation, through new organizational structures such as the interdisciplinary national laboratory, and through new experimental tool such as particle accelerators and nuclear reactors (Leslie 1993, p 64). Before World War II nuclear physics had not been a major focus of research in physics in the United States, but World War II and the Cold War focused new attention on the field. The U.S. Department of Defense realized that advances in nuclear weapons and nuclear reactors would require new fundamental knowledge as well as knowledge about actual bombs and reactors (Leslie 1993, p 174; Schweber 1992). At the same time, universities began to realize that the government, particularly the military, could provide needed support and funding if new research areas could lead to practical weapon systems. A number of American universities attempted to replicate the wartime approach to nuclear research by establishing laboratories, or centers, modeled after the Manhattan Project laboratory at Los Alamos, New Mexico (Schweber 1992, p 178). With support from the Office of Naval Research, MIT established the Laboratory for Nuclear Studies in 1945 which was followed by Cornell University's Laboratory of Nuclear Studies and Stanford University's High Energy Physics Laboratory (Galison et al. 1992, p 63; Schweber 1992, pp 177–178). Leslie has argued that this governmental support changed the way physics was done (Leslie 1993, p 134). Unlike science in Europe before World War II, research into nuclear physics in America after World War II combined a theoretical and an experimental approach and since the basic experimental approach required the building and operating of large scale particle accelerators and nuclear reactors, the distinction between physics and engineering often became blurred.

5.3 Big Science

The second development that arose from the new relationship between science, technology and the military was the rise of big science, a term coined by Alvin Weinberg and popularized by historian Derek J. de Solla Price in the 1960s to refer to the dramatic scale and complexity of scientific projects brought about by access to government funding (Weinberg 1961; Price 1963). Many scholars point to the Manhattan Project, which produced the atomic bomb, high energy physics experiments, the space program, and the human genome project, as examples of big science. Such projects required not only millions of dollars but thousands of researchers and technicians. The term big technology might equally apply to such research since an essential component of projects labeled as big science was the development of new large scale, complex, and expensive technologies, such as nuclear reactors, particle accelerators, bubble chambers, rockets, satellites, space telescopes and high speed computers, which were needed to carry out big science research. As such, big science, like the military-industrial-academic complex, created a new interdependent relationship between science and technology, establishing what might be called military-based science that again made science dependent upon technology. Philosopher Stephen Toulmin noted that after World War II the basic focus of scientific research was no longer nature herself but some unit of technology such as the nuclear reactor, the missile, or the computer (Toulmin 1964; Capshew and Rader 1992, p 9).

Peter Galison has shown how high energy particle physics has been shaped by the technology of particle detectors (Galison 1997, pp 19–31; Galison 1989). He has argued that two major experimental traditions arose in particle physics research in the twentieth century. The first approach, which he labels the image tradition, sought to capture a visual image of an event through cloud chambers and bubble chambers, and it focused on the discovery of a single "golden event." The opposing approach, which he labels the logic tradition, sought to make discoveries through statistical arguments based on amassing a large number of examples through Geiger counters or spark detectors. Each approach had its own strengths and weaknesses. While the image approach provided a detailed description of a particular particle interaction, there was always the risk that a particular golden event was an anomaly. On the other hand the logic tradition provided statistical evidence that a particular particle or event actually existed but it lacked the ability to provide a detailed description of those events.

Often in the big science of high energy particle physics the close relationship between theory and experiment led to a blurring of the distinctions between the two and transformed pure science into something closer to engineering science. Often experiments discovered particles that confirmed new theories and many times the main justification for building or improving an accelerator was to test a specific theory. Galison has argued that because of the immensely complicated technology that is at the basis of many of the newest particle detectors and the difficulty in obtaining straightforward predictions from abstruse theories such as quantum chromodynamics or string theory, "the boundary between theory and experiment became less sharp" (Galison 1997, p 43) As a result, many of the big science accelerator laboratories began to rely on "in house" theorists who interpreted theories or developed their own theories that were much more directly linked to a specific machine. At the same time, theorists outside the large experimental laboratories began to develop theories that could only be specifically tested with some particular experimental apparatus.

Although big science projects led to new fundamental and basic knowledge about the natural world, in most cases that new knowledge was intimately interconnected to technological knowledge of some humanly constructed experimental equipment used to discover that new knowledge. Also, while much of big science was focused on basic research, elements of that research were often closely related to military needs. Nuclear and particle physics held out the promise of more powerful weapons; particle detectors could be useful in monitoring an enemy's nuclear tests; the technology of accelerators might lead to new electronics, improved radar or even accelerated beam weapons; satellites, space stations, and space telescopes could lead to intelligence gathering, and improved rockets had obvious military applications. All this has led some scholars to talk about a period of post-academic science (Ziman 1996).

6 Conclusions

During the period of the Cold War, developments associated with the military-industrial-academic complex and developments of big science began to come together to form what has been called technoscience. Here distinctions between science and technology have not only blurred they have often completely disappeared. This is especially true in the case of the emergence of such new fields as artificial intelligence, computer science, virtual reality, synthetic life and genetic engineering. Even the names of these fields combine words that we associate with the scientific and natural world, such as intelligence, science, reality, life, and genetic, with words more closely associated with the technological world, such as artificial, computer, virtual, synthetic, and engineering (Gibbs 2004). A common element in many new fields of scientific thought is the idea of computation, but this has further blurred the distinctions between science and technology. Computation could be seen as either a human construction and therefore technological or as a branch of mathematics and therefore a science.

During the second half of the twentieth century scientists discovered that they could use the computer not only to solve scientific problems but that the computer could serve as a new model for understanding scientific phenomena (Edwards 1996). The computer has served as a model for understanding the human mind as well as the functioning of DNA and genetics (Alon 2006; Gardner 1985). But recently some scientists have begun to argue that the universe at its base is essentially computational and can be best understood in terms of a computer. A typical digital computer represents information or data as a string of 0s and 1s, or bits. Computation or information processing simply involves changing some bits from 0s to 1s, 1s to 0s, or leaving them unchanged. Modern quantum mechanics has a similar binary aspect in which a particle, like an electron, when measured or observed, can exist at a certain location or state, or not exist at that location or state. This has led some scientists, such as theoretical physicist John Archibald Wheeler, to argue that the entire universe is the result of a series of yes or no choices that take place when measurements or observations are made at the quantum level and can be summarized by the phrase "it from bit" (Wheeler 1998, pp 340–341).

In an article in *Physical Review Letters*, Seth Lloyd, a pioneer in quantum computing, argued that the universe could be understood as functioning in terms of information processing (Lloyd 2002). He argued that the state of every piece of matter in the universe could also represent the storage of information in the same way a set of coins could store information by being a series of heads and tails. The motion, or changing states of those pieces of matter, like the flipping of some of the coins, could then represent information processing. This would mean that the motions of all of the particles in the universe could be simply a form of computation or information processing. Since actual computers are composed of matter that is governed by the same physical laws that apply to the entire universe, Lloyd sees this as proof that at least a small part of the universe is capable of carrying out computation and information processing. So we see again the important role of technological thinking in science with the computer now replacing the heat engine, and before that the mechanical clock, as a model understanding the world (Channell 2004).

References

Ademann, H. B. (1966). *Marcello Malpighi and the evolution of embryology* (Vol. 1). Ithaca: Cornell University Press.

Alon, U. (2006). *An introduction to systems biology*. Boca Raton: Chapman & Hall/CRC.

Archimedes. (1909). *Geometrical solutions derived from mechanics*. Chicago: Open Court.

Bacon, F. (1937). In R. Jones (Ed.), *Essays, advancement of learning, new Atlantis and other pieces*. New York: Odyssey Press.

Basalla, G. (1962). William Harvey and the heart as a pump. *Bulletin of the History of Medicine, 36*, 467–470.

Belloni, L. (1975). Marcello Malpighi and the founding of anatomical microscopy. In M. L. Righini Bonelli & W. R. Shea (Eds.), *Reason, experiment and mysticism in the scientific revolution*. New York: Science History Publications.

Boas, M. (1952). The establishment of the mechanical philosophy. *Osiris, 10*, 412–541.

Böhme, G., Van den Daele, W., & Krohn, W. (1978). The scientification of technology. In W. Krohn (Ed.), *Dynamics of science and technology*. Dordrecht: D. Reidel.

Briggs, J. (1989). *Francis Bacon and the rhetoric of nature*. Cambridge: Harvard University Press.

Brush, S. (1967). Thermodynamics and history. *The Graduate Journal, 7*, 477–565.

Bush, V. (1945). *Science, the endless frontier*. Washington, DC: United States Government Printing Office.

Cantor, G., & Hodge, M. (Eds.). (1981). *Conception of the ether: Studies in the history of ether theories 1740–1900*. Cambridge: Cambridge University Press.

Capshew, J., & Rader, K. (1992). Big science: Price to present. *Osiris, 7*, 3–25.

Cardwell, D. (1971). *From Watt to Clausius: The rise of thermodynamics in the early industrial age*. Ithaca: Cornell University Press.

Cardwell, D. (1972). Science and the steam engine, 1790–1825. In P. Mathias (Ed.), *Science and society, 1600–1900*. Cambridge: Cambridge University Press.

Cardwell, D. (1976). Science and technology: The work of James Prescott Joule. *Technology and Culture, 17*, 674–686.

Cardwell, D. (1995). *The Norton history of technology*. New York: W.W. Norton.

Channell, D. (1989). *The history of engineering science: An annotated bibliography*. New York: Garland.

Channell, D. (1991). *The vital machine: A study of technology and organic life*. New York: Oxford University Press.

Channell, D. (2004). The computer at nature's core. *Wired Magazine, 2*, 79–80.

Channell, D. (2009). The emergence of the engineering sciences: A historical analysis. In A. Meijers (Ed.), *Handbook of the philosophy of technology and engineering sciences* (pp. 117–154). Amsterdam: Elsevier Science.

Cipolla, C. (1978). *Clocks and culture: 1300–1700*. New York: W.W. Norton.

Drachmann, A. G. (1948). *Ktesibios, Philon and Heron: A study in ancient pneumatics*. Copenhagen: Ejnar Munksgaard.

Drachmann, A. G. (1962). *The mechanical technology of Greek and Roman antiquity*. Madison: University of Wisconsin Press.

Edwards, P. (1996). *The closed world: Computers and the politics of discourse in cold war America*. Cambridge: MIT Press.

Forman, P. (2007). The primacy of science in modernity, or technology in postmodernity, and of ideology in the history of technology. *History and Technology, 23*, 1–152.

Fox, R., & Guagnini, A. (1998–1999). Laboratories, workshops, and sites: Concepts and practices of research in industrial Europe, 1800–1914. *Historical Studies in the Physical and Biological Sciences, 29*, 55–140, 191–294.

Galilei, G. (1954). *Dialogues concerning two new sciences*. New York: Dover Publications.

Galison, P. (1989). Bubbles, sparks, and the postwar laboratory. In L. Brown, M. Dresden, & L. Hoddeson (Eds.), *Pions to quarks: Particle physics in the 1950s* (pp. 213–251). Cambridge: Cambridge University Press.

Galison, P. (1997). *Image and logic: A material culture of microphysics*. Chicago: University of Chicago Press.

Galison, P., Hevly, B., & Lowen, R. (1992). Controlling the monster: Stanford and the growth of physics research, 1935–1962. In P. Galison & B. Hevly (Eds.), *Big science: The growth of large-scale research*. Stanford: Stanford University Press.

Gardner, H. (1985). *The mind's new science: A history of the cognitive revolution*. New York: Basic Books.

Gibbs, W. W. (2004). Synthetic life. *Scientific American, 290*(5), 74–81.

Hall, A. R. (1959). The scholar and the craftsman in the scientific revolution. In M. Clagett (Ed.), *Critical problems in the history of science*. Madison: University of Wisconsin Press.

Hall, M. B. (Ed.). (1973). *The Pneumatics of Hero of Alexandria: A facsimile of the 1851 Woodcroft edition*. New York: American Elsevier Publishing Co.

Hankins, T., & Silverman, R. (1995). *Instruments and the imagination*. Princeton: Princeton University Press.

Harman, P. (1982). *Energy, force and matter: The conceptual development of nineteenth-century physics*. Cambridge: Harvard University Press.

Healey, E. C. (1856). To our readers. *The Engineer, 1*, 3.

Heidegger, M. (1977). *The question concerning technology and other essays*. New York: Harper and Row.

Hills, R. (1989). *Power from steam: A history of the stationary steam engine*. Cambridge: Cambridge University Press.

Hobbes, T. (1839). *The English works of Thomas Hobbes* (Vol. 7). London: John Bohn.

Hodges, H. (1970). *Technology in the ancient world*. New York: Barnes and Noble Books.

Holton, G. (1973). *Thematic origins of scientific thought: Kepler to Einstein*. Cambridge: Harvard University Press.

Ihde, D. (1990). *Technology and the lifeworld: From garden to earth*. Bloomington: Indiana University Press.

Ihde, D. (2010). *Heidegger's technologies: Postphenomenological perspectives*. New York: Fordham University Press.

Kay, L. (2000). *Who wrote the book of life: A history of the genetic code*. Stanford: Stanford University Press.

König, W. (1996). Science-based industry or industry-based science? Electrical engineering in Germany before World War II. *Technology and Culture, 37*, 70–101.

Koyré, A., & Cohen, I. B. (1962). Newton and the Leibniz-Clarke Correspondence. *Archives Internationales d'Histoire des Sciences, 15*, 63–126.

Kuhn, T. (1969). Energy conservation as an example of simultaneous discovery. In M. Clagett (Ed.), *Critical problems in the history of science*. Madison: University of Wisconsin Press.

Kuhn, T. (1978). *Black-body theory: The historical development of quantum theory and quantum discontinuity, 1894–1912*. New York: Oxford University Press.

Landels, J. G. (1978). *Engineering in the ancient world*. Berkeley: University of California Press.

Landes, D. (1983). *Revolution in time: Clocks and the making of the modern world*. Cambridge: Harvard University Press.

Latour, B. (1987). *Science in action: How to follow scientists and engineers through society*. Cambridge: Harvard University Press.

Leslie, S. W. (1993). *The cold war and American science: The military-industrial-academic complex at MIT and Stanford*. New York: Columbia University Press.

Lloyd, S. (2002). Computational capacity of the universe. *Physical Review Letters, 88*(23), art. no. 237901.

Martin, J. (1992). *Francis Bacon, the state and the reform of natural philosophy*. Cambridge: Cambridge University Press.

Mayr, O. (1986). *Authority, liberty & automatic machinery in early modern Europe.* Baltimore: The Johns Hopkins University Press.

McMullin, E. (1990). Conceptions of science in the scientific revolution. In D. Lindberg & R. Westman (Eds.), *Reappraisals of the scientific revolution.* Cambridge: Cambridge University Press.

Meyer-Thurow, G. (1982). The industrialization of invention: A case study from the German dye industry. *Isis, 73,* 363–381.

Mumford, L. (1963). *Technics and civilization.* New York: Harcourt, Brace, and World.

Noble, D. (1997). *Religion of technology: The divinity of man and the spirit of invention.* New York: Knopf.

Pagel, W. (1967). *William Harvey's biological ideas.* New York: Hafner.

Price, D. d. S. (1963). *Little science, big science.* New York: Columbia University Press.

Reich, L. (1985). *The making of American industrial research: Science and business at GE and Bell, 1876–1926.* Cambridge: Cambridge University Press.

Roland, A. (1985). Technology and war: A bibliographic essay. In M. R. Smith (Ed.), *Military enterprise and technological change.* Cambridge: MIT Press.

Roland, A. (2001). *The military-industrial complex.* New York: Society for the history of technology and the American Historical Association.

Rosenfield, L. (1968). *From beast-machine to man-machine: Animal soul in French letters from Descartes to La Mettrie.* New York: Octagon.

Salomon, J.-J. (1973). *Science and politics.* Cambridge: MIT Press.

Sassower, R. (1995). *Cultural collisions postmodern technoscience.* New York: Routledge.

Schweber, S. (1992). Big science in context: Cornell and MIT. In P. Galison & B. Hevly (Eds.), *Big science: The growth of large-scale research* (pp. 149–183). Stanford: Stanford University Press.

Shapin, S. (1981). Of gods and kings: Natural philosophy and politics in the Leibniz-Clarke disputes. *Isis, 72,* 187–215.

Shapin, S. (1996). *The scientific revolution.* Chicago: University of Chicago Press.

Shapin, S., & Schaffer, S. (1985). *Leviathan and the air-pump: Hobbes, Boyle, and the experimental life.* Princeton: Princeton University Press.

Smith, C. (1998). *The science of energy: A cultural history of energy physics in Victorian Britain.* Chicago: University of Chicago Press.

Staudenmaier, J. (1985). *Technology's storytellers: Reweaving the human fabric.* Cambridge: MIT Press.

Toulmin, S. (1964). The complexity of scientific choice: A stocktaking. *Minerva, 2,* 343–359.

Webster, C. (1965). William Harvey's conception of the heart as a pump. *Bulletin of the History of Medicine, 39,* 510–515.

Weinberg, A. (1961). Impact of large-scale science. *Science, 134,* 161–164.

Wheeler, J. A. (1998). *Geons, black holes and quantum foam: A life in physics.* New York: W.W. Norton.

Wise, M. N. (1990). Electromagnetic theory in the nineteenth century. In R. C. Olby et al. (Eds.), *Companion to the history of modern science* (pp. 342–356). London: Routledge.

Zagorin, P. (1998). *Francis Bacon.* Princeton: Princeton University Press.

Ziman, J. (1996). Post-academic science: Constructing knowledge with networks and norms. *Science Studies, 9,* 67–80.

Chapter 4
The Scientific Use of Technological Instruments

Mieke Boon

Abstract One of the most obvious ways in which the natural sciences depend on technology is through the use of instruments. This chapter presents a philosophical analysis of the role of technological instruments in science. Two roles of technological instruments in scientific practices are distinguished: their role in discovering and proving scientific theories, and their role in generating and investigating new physical phenomena that are of technological relevance. Most of the philosophy of science is theory-oriented and therefore tends to ignore the importance of producing and investigating physical phenomena in current scientific practices. This chapter selectively chooses some recent trends in the philosophy of science that relate to the role of technological instruments in order to indicate the potential for philosophical accounts of scientific practices that productively integrate the two roles of technological instruments.

1 Introduction

At present, many accept that modern science and technology are interwoven into a complex that is sometimes called 'technoscience'.[1] When focussing on how technology has an effect on scientific practices, two roles can be distinguished. On

---

[1] The focus of this article will be on techno-science in the narrow sense, namely, how experimentation and instrumentation in scientific practices is entangled with the development of technological instruments – which, on the one hand, have a role to play in scientific practices, but on the other hand, will be developed further into concrete technological devices outside the laboratory. A focus on experimentation and instrumentation as the crucial link between science and technology draws attention to materiality, such as technologically relevant physical phenomena, which are produced through experimentation and instrumentation. Conversely, most of the traditional philosophy of science focuses on theories as the major aim of science. Techno-science as a movement addresses much broader issues, which are not commonly addressed in the philosophy of science but which are part of the philosophy of technology (see, for instance, Idhe and Selinger 2003, and Radder 2004).

M. Boon (✉)
Department of Philosophy, University of Twente, PO Box 217, 7500 AE Enschede, The Netherlands
e-mail: m.boon@utwente.nl

the one hand, much of the progress of science is driven by the sophistication of instrumentation. Conversely, scientific research plays a crucial role in the development of *high-tech* devices. Related to these two roles of technology in science, two functions of technological instruments can be distinguished: Firstly, as experimental equipment and measuring devices, which have an *epistemological* function in developing and testing scientific theories and models. Secondly, technological devices play a *material* role in the production of specific physical phenomena, in particular those that may be of interest for performing a technological function.

This is exemplified by a simple historical example such as Robert Hooke's experimental set-up in which the elasticity of a spring was studied. On the one hand, the experimental set-up of springs and weights and measurements of lengths has an epistemological function in finding laws that describe this phenomenon, such as Hooke's law. At the same time, the physical phenomenon – the elasticity of a device or material – is utilized in technological applications. Scientific research in this application context aims, for instance, at an understanding of the elastic behaviour such that it can be produced or manipulated in specific objects, materials or devices. Another example is superconductivity in mercury at temperatures near absolute zero, which is a phenomenon that on the one hand calls for a scientific explanation, while on the other hand the technological significance of this phenomenon calls for scientific research that demonstrates how this phenomenon can be produced or manipulated under different conditions and in different materials. Some scientific research projects, for instance, are dedicated to the development of ceramic materials that are superconductive at higher temperatures, a feature which is crucial for viable technological applications.

These two roles of technological instruments in science can be captured in terms of two distinct philosophical perspectives on modern scientific research. Firstly, the traditional view of the philosophy of science in which the focus is on the theories. From this perspective, the primary aim of science is the discovery and justification of theories, for which the development of technological instruments is subservient. The alternative perspective considers scientific research in application contexts. Broadly speaking, it focusses on the technological production and measurement of physical phenomena significant to technological applications in the broad sense, which includes technology in the narrow sense, but also medicine and agriculture, and all kinds of measurement techniques. In this alternative view, scientific research

---

Alfred Nordmann (2012) presents a description of 'techno-scientific knowledge' similar to what I aim to say about the aim of scientific research from a 'phenomenon-oriented' (rather than 'theory-oriented') perspective on the role of technological instruments in scientific research: "Techno-scientific knowledge includes the acquisition and demonstration of basic capabilities. ... Rather than being applied science or applied techno-science, then, there is basic techno-scientific research which consists of demonstrated capabilities to visualize, to characterize substances, to measure and model – and, of course, to manipulate and control surprising phenomena" (Nordmann 2012, 19), also see Lacey (2012).

Also see Boon (2011) in which I present an overview of the literature on the relationships between science and technology, and an analysis of their epistemological relationship.

firstly aims at the material production and theoretical understanding, both of relevant phenomena as well as the technological instruments that create them. This presumes the simultaneous material and theoretical development of (1) relevant physical phenomena and the technological instruments that produce them, as well as (2) scientific knowledge that may be used to describe or explain both the phenomenon and the workings of the technological instruments – where 'scientific knowledge' includes empirical knowledge, theoretical concepts, scientific models, etc. Below, I will refer to these two perspectives on science as 'theory-oriented' and 'phenomena-oriented'.

The 'theory-oriented' perspective has been dominant in traditional philosophy of science. In most of that tradition, the role of technological instruments has been neglected. Only in the last few decades has progress been made towards a philosophical understanding of the epistemological roles of instruments and experiments in the justification of scientific theories. This chapter aims to give an overview of this important movement in the philosophy of science.[2]

The structure of this chapter is as follows: Sect. 2 will sketch the philosophical background against which interest in the role of technological instruments in science emerged. Sects. 4, 5, and 6 present an outline of so-called *New Experimentalism* and other trends in the philosophy of science that address the role of experiments and technological instruments in scientific research, focussing on: (1) philosophical accounts of their roles in the justification of scientific theories, and (2) taxonomies and epistemologies of instruments and experiments. These topics will be presented within the 'theory-oriented' perspective from which science tackles the discovery and justification of theories. But it will become clear along the way that these topics are also relevant to a better understanding of the second aim of modern science, namely, the invention of 'high-tech' things – such as instruments, materials, and apparatus – which relies on a scientific understanding of phenomena that are technologically produced and the sophisticated instruments used in their creation. In Sect. 7, it will be argued that a full-grown philosophical account of this second, epistemic and material role of technological instruments in scientific research is still lacking. Nonetheless, the New Experimentalist movement, together with some recent work in the philosophy of science that explains *how* the development of technological instruments and the formation of theoretical concepts are crucially entangled, give directions towards a more viable understanding of both the character of scientific knowledge and of how scientific research contributes to the development of high-tech devices – on this track, the two perspectives merge in a productive manner.

---

[2] Authors such as Don Ihde (1991, 2009), Hans Radder (1996, 2003a), Davis Baird (2003, 2004), Joe Pitt (2000) and several others have been pioneering in attempts to build a bridge between the philosophy of science and the philosophy of technology by emphasizing the epistemic and material role of instruments in scientific practices, and trying to steer away from a theory-oriented perspective on science. Although I acknowledge the significance of their contributions, I will focus on developments towards such insights within 'main stream' philosophy of science.

2 Positivistic Philosophy of Science

In a traditional positivistic view, the aim of science is the production of reliable, adequate and/or true knowledge about the world. Positivistic philosophy of science thus focuses on theories produced in science. Its task is to produce accounts of confirmation and inductive inference that justify these theories. It assumes that the role of instruments and experiments is for testing hypotheses in controlled laboratory settings, and as such the instrumentation and experimentation are mere data providers for the evaluation of theories. Positivists do not make a distinction between observation through our senses and observation by means of instruments in experiments. Instruments are *instrumental* to the articulation and justification of scientific knowledge of the world. Although not doing full justice to the positivists view, the implicitly held metaphor is that scientists observe nature through technological spectacles that are not thought to significantly influence the resulting picture of nature (also see Rouse 1987[3]). As a consequence, the ways in which pictures and data are produced and evaluated by means of instruments and experiments has not been a topic of much concern to positivistic philosophy of science.

A classical problem for the positivistic idea of testing theories is the Duhem-Quine problem of under-determination of theories by empirical evidence. If an experiment or observation is persistently inconsistent with theory, one could either revise the theory, or revise the auxiliary hypotheses – for instance those that relate to the proper functioning of the instruments. An additional severe attack to the positivistic image of science came from Popper (1959, 156), who claims that all observation is theory-laden. To him, observations, and even more so observation statements and statements of experimental results, are always interpretations of the facts observed; they are *interpretations in the light of theories* (also see Van Fraassen 2012). Kuhn's notion of paradigms was conceived in a similar vein: rather than observation, the paradigm is basic, and observations only exist insofar as they emerge within the paradigm.

Since non-empirical factors seem to dominate the experimental work from its initial planning to its final result, the philosophical idea that theories are tested in experiments has become problematic. This is a severe threat to philosophical justification of the empirical and logical methodology of testing scientific theories proposed in logical positivism and empiricism. Signifying and addressing these threats has been very influential because it opened the road to extreme sceptical appraisals of science.

---

[3] According to Rouse (1987): One might say that the traditional philosophical model of the local site of research is the observatory. Scientists look out at the world to bring parts of it inside to observe them. Whatever manipulative activities they perform either are directed at their instruments or are attempts to *re*produce phenomena in a location and setting where they will be observable. The new empiricism leads us instead to take seriously the *labor*atory. Scientists produce phenomena: many of the things they study are not 'natural' events but are very much the result of artifice (Rouse 1987, 23).

Recent approaches in the philosophy of science have changed focus from the logic of justification of scientific theories to the role of experiments and instruments in scientific reasoning. These new approaches can be understood as an attempt to solve two problems simultaneously. On the one hand, they aim to shed light on the problem of whether science can test theories if experimental results involve theoretical interpretations. On the other hand, they aim at resisting the scepticism that resulted from embracing the theses of under-determination of theory by evidence and the theory-ladeness of observation. They suggest, instead, that understanding the problems of under-determination and theory-ladeness involves understanding the role of experiments and instruments in testing scientific theories. Philosophical issues raised by these recent approaches include: How can experimental results be justified, that is, how do we know whether we observe a natural phenomenon or rather an artifact of the experimental set-up. How do we know that the instrument is functioning properly? Can we observe nature by means of our instruments without influencing the resulting picture of nature if the design of these instruments involves theoretical considerations? Can we make an ontological distinction between *Nature* and instrument anyway? Do theories confirmed in the laboratory apply to the world outside, and how can we generalize experimental results if at all? These are the kind of questions that are of interest to the so-called *New Experimentalists*.

3 New Experimentalism

Robert Ackermann gave the name of *New Experimentalism* to philosophical discussions that focused on the role of experiments and instruments in science, and which therefore, attempted to overcome the draw-backs of logical empiricism, and so avoid the scepticism of the social constructivists (Ackermann 1985, 1989; also see Mayo 1994). Some of the key figures of this movement from the 1980s and early 1990s included Ian Hacking, Nancy Cartwright, Allan Franklin, Peter Galison, Ronald Giere, Robert Ackermann himself, and more recently, Deborah Mayo.

Although each of these authors has a different focus, they share several viewpoints on the course of philosophy of science. They defend a philosophy that considers scientific practices, and they do not accept the restriction to the logic of science that positivistic philosophers had set for themselves. Traditional philosophical accounts of how observation provides an objective basis for evaluation of theories by the use of confirmation theory, or inductive logic, should be replaced by accounts of science that reflect how experimental knowledge is actually arrived at and how this knowledge functions. Obviously, the traditional distinction between the context of discovery and the context of justification, which motivated philosophers to restrict their task to the logic of justification of scientific theories, is abandoned. Instead, the New Experimentalists aim at an account of the rationality of scientists in scientific practices that includes how scientists reason about experiments, instruments, data, and theoretical knowledge.

Philosophical progress in this new tradition leans heavily on historical case studies that focus on aspects of experiments and instruments. These historically informed approaches strengthen the tradition that may have been ushered in by Thomas Kuhn, and which is now called the 'history and philosophy of science'. Reliance on historical case studies does not necessarily open the door to sociologically flavoured explanations of science, since an additional viewpoint shared by the New Experimentalists is that abstracting from sociological aspects of scientific practices, i.e. focussing on elements internal to scientific practice, is justified. Thus, the focus is on epistemological aspects of experiments, instruments, data and the processing of data, and different layers of theorizing, rather than the relationships between scientists, instrument builders, laboratories, editors, journalists, industry, government, media, and the public, which are important to social studies of science. Although, these philosophers admit that sociological and contingent factors may determine the course of science, they deny that sociological factors are determining methodological and epistemological criteria internal to scientific practices.

In general, the New Experimentalists share the view that focusing on aspects of experiments and instruments holds the key to avoiding or solving a number of problems, such as the under-determination of theory by empirical knowledge, the theory-ladeness of observation, and extreme sceptical positions – such as social constructivist – that result from it. In their view, these problems stem from the theory-dominated perspective on science of positivistic philosophers of science.

4 Theory-Ladeness of Observation

4.1 *'Experiments Have a Life of Their Own'*

In 1913 the Dutch physicist Heike Kamerlingh Onnes was able to achieve in his experiments temperatures near to absolute zero, and he discovered that mercury became superconductive under these circumstances. It took to 1957 before Bardeen, Cooper, and Schrieffer put forward a theoretical explanation of this phenomenon – the BCS theory. These kinds of discoveries of new phenomena can be taken as counter-examples to Popper's claim that observation in an experiment is always a theoretical interpretation of the facts observed. Some phenomena are observed for which there is no theory. The discovery of superconductivity resulted from studying the behaviour of metals at temperatures near absolute zero. This discovery was independent of the theory that later explained the phenomenon. In scientific practice, some experiments produce reproducible phenomena – often called 'phenomenological laws' – that await a theoretical explanation. Therefore, observations of experimentally produced phenomena such as superconductivity, are not always theory-laden in a problematic sense.

Ian Hacking (1983) used these kinds of historical cases of discoveries of new phenomena to argue that 'experiments have a life of their own', meaning experiments have more objectives than the mere aim of *testing* theories (also see: Galison 1987, 1997; Steinle 2002, 2006; Radder 2003a, b, and Franklin 2005). Hacking gives several other examples in which observations of phenomena were possible before there was any theory to explain them. The shift of emphasis on the role of experiments for testing theories to their role in producing reproducible phenomena, as indicated by Hacking, also indicates a shift of focus in the philosophy of science regarding the aim of science.

The important contribution of the discovery of superconductivity was not that it confirmed a theory about the world, but the discovery of that phenomenon, i.e. that such a phenomenon exists (it is worth keeping in mind that this phenomenon is not naturally occurring, but technologically produced). Furthermore, aiming at a theoretical understanding of this phenomenon and of the materials and physical conditions that produce it is not just driven by scientific curiosity, but also by the goal of technologically producing and utilizing this physical phenomenon.

4.2 'Data' Versus 'Meaning of Data'

In a strict positivistic view, one could refuse to acknowledge that *observing* superconductivity of metals near absolute zero involves theories of the instruments that are used for measuring the conductivity of the metal. This is one of the topics addressed by Ackermann (1985). He agrees that *data* given by instruments – such as data produced by a conductivity meter – may be given independent of theory. Instruments create an invariant relationship between their operations and the world, at least when we abstract from the expertise involved in their correct use. These readings are also independent of changes in the theory. An instrument reads 2 when exposed to some phenomenon. After a change in theory, it will continue to show the same reading, but we may take the reading to be no longer important, or, to tell us something other than what we thought originally. Thus, the *meanings of data* – such as superconductivity – are not given by the data, since the data are interpreted as such and such phenomenon through the use of theories. Without a theory and theoretical expectations, some data may not even be noticed. In addition, data may be interpreted differently by different theories (also see Van Fraassen 2012). When our theories change, we may conceive of the significance of the instrument and the world with which it is interacting differently, and the datum of an instrument may change in significance, but the data can nonetheless stay the same, and will typically be expected to do so. However, according to Ackermann, although data have an internal stability, which is reproducible through the use of instruments, their meaning is neither manifest nor stable.

Also other authors have criticized the idea that observations by means of instrument are independent of theory. Hacking's (1983) idea has been criticized for having a positivistic stance towards experiments very similar to Logical Empiricism (e.g., Carrier 1998). Others have argued that experiments often do not reveal actual states of affairs. In exploratory experiments it requires the formation of new basic concepts – such as the notion of a current circuit in the case of Ampère – before the data produced by the instrument can be interpreted as a phenomenon (e.g., Harré 1998; Steinle 2002, 2006).

One of the issues at stake is how new theoretical concepts (such as superconductivity) are formed in cases such as the experiments by Kamerling Onnes. Since Popper (1959), and Kuhn (1970), a common reply in the philosophy of science has been that the formation of new theoretical concepts on the basis of experimental observation involves the theory that is supposed to be tested or discovered by these observations (i.e., the problem of the theory-ladeness of observation). Traditionally this entanglement of theories and observation was considered problematic as it threatens the objectivity of science. In particular, recent approaches in the philosophy of science suggest that the formation of new *theoretical concepts* for interpreting experimental observations – such as those made by Kamerling Onnes – are entangled with the *formation* of theories, on the one side, and the development of instruments and experiments, on the other, in a process called 'triangulation' (also see below, Hacking 1992). I will argue that these enriched accounts of the relationship between theories and experimental observations agree with the 'phenomena-oriented' perspective on the role of technological instruments in scientific research. This opens the way for a philosophy of science that explains how scientific research contributes to the development of technological devices.

4.3 'Representing' Versus 'Intervening'

An important claim by Hacking (1983) is that much of our empirical knowledge does not result from passive observation by means of instruments, but from *interventions* with instruments. The spectacle metaphor of instruments is replaced by a metaphor where the instruments become a material playground that provides us with a way to learn a lot about the world and about phenomena produced by these instruments. Observation as a source of empirical knowledge is extended to include *doing*, by *interacting* and *intervening* with the world through our instruments. Only by intervening do we discover the material resistances of the world. This source of empirical knowledge provides additional constraints that may overcome the problem of under-determination of theory by observation (also see Sect. 5.1).

Hacking's emphasis on the role of intervention is an important step towards an orientation on the role of phenomena in scientific research. By interacting and intervening with technological instruments, scientists firstly produce (new) phenomena. As has been indicated in the former section, a phenomenon is established through the entangled activities of materially producing it and forming a theoretical concept. Next, scientists aim to reproduce or manipulate the phenomenon through interventions with instruments and experimental set-ups, which enable investigation, but at the same time generate and shape the phenomenon.

Following up on Hacking's notion, it will be argued (in Sect. 7) that according to a phenomenon-oriented perspective, instead of accurate representations of the world as it is being the aim of scientific research, the researchers, often try to understand the results of active interventions with the world by means of technological instruments.

5 Underdetermination of Theory by Empirical Data

5.1 *The Self-Vindication of the Laboratory Sciences*

One way to avoid the Duhem-Quine problem of underdetermination has been proposed by Hacking (1992). He claims that our preserved theories and the world fit together, not solely because we have found out how the world is, but rather we have tailored each to the other. As laboratory science matures, it develops a body of types of theory and types of apparatus and types of analysis of data that are mutually adjusted to each other. Any test of theory is related to apparatus that has evolved in conjunction with it – and in conjunction with modes of data analysis. Conversely, the criteria for the working of the apparatus and for the correctness of analyses are precisely the fit with theory. For instance, phenomena are not described directly by Newtonian concepts. It is rather certain measurements of the phenomena – generated by a certain class of what might be called 'Newtonian instruments' – that mesh with Newtonian concepts. The accuracy of the mechanics and the accuracy of the instruments are correlative. This process of tailoring the elements of experiments to fit together is what Hacking calls the self-vindication of laboratory science.

To explain this idea, Hacking proposed a taxonomy of experiments, which consists of 15 elements internal to the experiments. This list of elements is divided into three groups, which are: (1) the intellectual components of an experiment ('ideas'), such as systematic theories, background knowledge, theoretical models of the apparatus, phenomenological laws, and hypotheses; (2) the material substance that we investigate or with which we investigate ('things'), such as the apparatus, the tools and instruments, the substances, and the material objects investigated; and (3) the outcomes of an experiment and the subsequent manipulation of data, such as data-reduction, calculations that produce more data, and interpretations ('marks').

Contrary to the Duhem-Quine thesis that theory is underdetermined by data, Hacking argues that the constraints of these 15 elements allow too few degrees of freedom. All the elements can be modified, but when each one is adjusted with the others so that our data, our machines, and our thoughts cohere, interfering with any one throws all the others out of kilter. It is extraordinarily difficult to make one coherent account, and it is perhaps beyond our powers to make several (Hacking 1992, 55).

Next, Hacking argues for a new conception of how theories are tested in experiments. This conception disagrees with positivistic ideas of testing theories. Theories are not checked by comparison with a passive world with which we hope they correspond, but with a world in which we intervene. We do not formulate conjectures and then just look to see if they are true. Instead, according to Hacking, we invent devices that produce data and isolate or create phenomena, and a network of different levels of theory is true to these phenomena. Conversely, we may in the end count them as phenomena only when the data can be interpreted by theory. Thus there evolves a curious tailor-made fit between our ideas, our technological instruments, and our observations, which Hacking calls a coherence theory of thought, action, material things, and marks.

In this approach, Hacking shows another way of dealing with the problems of the theory-ladeness of observation, the under-determination of theory by empirical data, and the scepticism to which this could lead. Instead of showing – by means of case-studies – that many exceptions to these problems can be found in scientific practice, and thus, that the philosophical problems have been exaggerated, he now accepts that theories, things and data are mutually adapted in order to create mutual coherence.

5.2 *Epistemology of Experiments and Instruments*

Allan Franklin (1986, 2002, 2005) addresses the question of how experimental results can be justified, that is, how we know that the instrument is functioning properly, and how we know whether we observe a natural phenomenon or a mere artefact of the instrument. In examining this problem he carried out detailed case studies of experiments in physics, and reconstructed how scientists reasoned about their results. From this analysis, he proposes a number of epistemological strategies that scientists regularly use, such as: observation of the same phenomena with different pieces of apparatus; prediction of what will be observed under specified circumstances; examination of regularities and properties of the phenomena themselves which suggest they are not artefacts; explanation of observations with an existing accepted theory of the phenomena; the elimination of all sources of error and alternative explanations of the results; evidence to show that there was no plausible malfunction of the instruments or background that would explain the observations; calibration and experimental checks; predictions of a lack of phenomena; and statistical validation. According to Franklin, in assessing their results, scientists act rationally when establishing the validity of an experimental result or observation on the basis of such strategies. He calls his articulation of the strategies of scientists who aim at validation of observation an *epistemology of experiment*. What is new in Franklin's approach to scientific reasoning is the crucial role played by 'errors'. In the validation of observations by means of instruments, he suggests, scientists focus on all possible sources of errors.

5.3 Error and the Growth of Experimental Knowledge: Learning from Error

Deborah Mayo (1996) is another author, who, in an attempt to explain how scientists approach the problem of reliable knowledge, emphasises the importance of learning from errors.

How do scientists know that empirical data provide a good test of, or reliable evidence for a scientific hypothesis? In a traditional approach, philosophers seek the answer in the logical relationship between evidence and hypotheses, producing theories of evidence, confirmation, and inductive inference. In contrast, Mayo claims that answering this question requires an account for what happens in the production of evidence in scientific practice. This calls for an analysis of how scientists acquire experimental knowledge, that is, of how the data were generated in specific experimental testing contexts.

In her view, scientists do not focus on evidence for their hypotheses. Instead, much of the effort is concentrated on the reduction of errors. This is done by articulating and testing alternative hypotheses, and by tracking down sources of errors in instruments and experiments. One of her core ideas is that scientists come up with hypotheses at many different epistemological levels of scientific experimentation, and put these hypotheses to *severe tests*. Thus, it is not the highbrow scientific hypothesis that philosophers are usually interested in that are put to test in the first place, but hypotheses about the proper functioning of instruments, hypotheses about the proper functioning of the object under study (e.g. does it require shielding to external influences), and hypotheses about the rightness of data that are usually inexact and 'noisy'. It is important to notice that this includes hypotheses on possible errors in the functioning of instruments and the system under study, and errors in the data produced. These hypotheses involve 'intermediate' theories of data, instruments, and experiment. It is in this manner that scientists track down errors, and correct data and make adaptations to the instruments and experimental set-up.

Mayo's key idea for specifying the character of 'severe tests of hypotheses' is the application of statistical tests. The severity of a test to which a hypothesis h is put, is related to the probability that a test would reject h if h were false. For instance, a test is a severe test of hypothesis h if h implies observation O, and there is a very low probability of observing O if h is false. In her view, the method of statistical tests is a means of learning. On this account, one learns through scientific inquiry, not from how much the evidence confirms the hypothesis tested, but rather from *discordant* evidence and from finding anomalies. Hence, scientists learn from 'error' and error correction, which is an important extension to Hacking's idea that they learn from *intervention*. Mayo's account shows how low-level testing works, and how this method of identifying sources and magnitudes of error, serves in testing higher-level theory. She thus shows that a piecemeal, 'bottom-up' analysis of how scientists acquire experimental knowledge yields important insights into how science works (Mayo 1996, 2000; also see Hon 1998 and Carrier 2001).

6 Theory-Ladeness of Instruments

6.1 Do We See Through a Microscope?

Additional to the problem of theory-ladeness of observation, there is the problem of whether we can observe nature by means of our instruments if the design of these instruments involves theoretical considerations. Several authors have given their support to the idea that the theory-ladeness problem of instruments can be excluded, at least in some of our observations with instruments (e.g., Baird 2003, 2004; Heidelberger 2003; Lange 2003; Rothbart 2003, see Sect. 6.4). A favoured example is observations by means of microscopes and other instruments with which objects can be made visible (e.g. Hacking 1983; Chalmers 2003).

Common ideas about the possibility of avoiding the problem that observations by means of instruments are theory-laden and therefore not objective, are reminiscent of a Lockean kind of empiricism, which supposes that a distinction can be made between primary and secondary properties. Primary properties (or features) of things are 'out there', in the world, while secondary properties are 'in us', in our minds. In Locke's time most measurement instruments that we know of today, did not exist. Therefore, the only primary properties of objects were: extension in space, shape, motion, number, and solidity (or impenetrability). These primary properties had the 'power' to cause *in us* not only perceptions of shape, motion, etc., but also perceptions of secondary properties, such as colour, sound, warmth or cold, odour, etc. These latter perceptions did not 'correspond' to the primary properties of material objects. Nowadays, many properties can be 'observed' by means of instruments and specific measurement procedures. Therefore, the list of supposed primary properties can be extended to include those that are measured; for instance, temperature, wave-length, electrical resistance, and magnetic field strength (also see the properties listed in *The Handbook of Chemistry and Physics*). The challenging question is how to understand the role of instruments in the measurement of these properties. Can they be called 'primary properties'? Or, differently put, in what sense do measurements by means of instruments *represent* properties of the world 'out there'?

6.2 Theory-Ladeness of Measurements

The epistemological ideal in the measurement of properties by means of instruments is technologically aided 'comparison' or 'observation,' or a combination of both. Metaphorically speaking, the epistemic contribution of a measuring instrument is similar, either to the use of an external standard such as the meter, or to the use of an instrument that enhances the senses such as a microscope or telescope. By means of these instruments certain properties or behaviours of the experimental set-up are 'compared' with a standard, or 'observed'. Ideally, the measured property or the phenomenon is isolated and quantified by the measuring instrument, but not in any significant sense physically or conceptually affected by it, let alone, produced by it.[4]

Hasok Chang (2004) who performed a detailed historical study into one of the most 'taken for granted' instruments of scientific practice – thermometers – has, however, challenged this idea. The development and testing of thermometers was radically different to the development of microscopes and telescopes, since the measurement of temperature lacked a solid reference. This led to an unavoidable circularity in the methodology of testing in thermometry. Scientists had to consider: How they could test whether the fluid in their thermometer expands regularly with increasing temperature without a circular reliance on the temperature-readings provided by the temperature itself. And, how they, without having established thermometers, could find out whether water boiled or ice melted always at the same temperature, so that those phenomena could be used as 'fixed points' for calibrating thermometers. Chang shows that the route to thermometers that gave correct temperature readings was long, and intellectually and experimentally challenging.

Although the so-called representational theory of measurement (e.g., Krantz et al. 1971; Suppes et al. 1990; Luce et al. 1990) assumes that theory-ladeness is not overly problematic to observations by means of measurement instruments, Chang shows that stabilizing observations by means of instruments involves many indispensable theoretical considerations. He arrives at a conclusion that is more constructivist than Hacking (1992) who takes as a basic idea that the material world warrants the objectivity of scientific knowledge and also, that empirical knowledge of the material world does not rely on our theories of it (also see Chang 2009b). In Chang's view, through philosophers' attempts to justify measurement methods we discover that circularity is inherent. His example of thermometry shows that finding empirical knowledge of temperature involved theoretical assumptions about the properties of matter and heat.

The basic problem for a philosophical account of empirical science is that it requires observations based on theories, whereas empiricist philosophy demands that those theories should be justified by observations. Chang holds that the only productive way of dealing with that circularity is to accept it and admit that justification in empirical science has to be *coherentist* instead of foundationalist. Within such coherentism, *epistemic iteration* provides an effective method of scientific progress, since it involves simultaneous corrections of interrelated theory (such as thermodynamics) and instruments (such as thermometers), which results in stable systems.

---

[4]Frigerio et al. (2010) present a clear outline of this so-called representational theory of measurement, which holds that to measure is to construct a representation of an empirical relational system to a numerical relational system, under the hypothesis that relations in the empirical relational system are somehow observable (Frigerio et al. 2010, 125). The crucial question regarding the theory-ladeness of measurement is to what extent the construction of a representation involves theory.

6.3 Nomological Machines

An important aspect revealed by Chang's analysis of the history of thermometry is the epistemic and technological efforts needed for constructing a stable technological device. It shows the entanglement of building a stable technological device (e.g., a thermometer), with 'stabilizing' empirical knowledge (e.g., that water boils at 100 °C), and producing theoretical knowledge (e.g., laws of thermodynamics). Understanding how a stable technological device is produced is relevant for a different take on a question such as: how the generalization of experimental results is justified, and why theories confirmed in the laboratory apply to the world outside. Returning to the musings of Hacking (1992), it will be argued in Sect. 7 that the stabilization of instruments such as described by Chang, is crucial for the stabilization and applicability of scientific knowledge.

Nancy Cartwright (1983, 1989, 1999) is one of the first authors who stressed the role of instruments in 'discovering' laws of nature. She holds that in the positivistic tradition, theoretical laws that are tested in experiments are conceived as necessary regular associations between properties. However, according to Cartwright, in order to test these laws we create so-called *nomological machines*. A nomological machine is a fixed arrangement of components, or factors, with stable capacities that in the right sort of stable environment will give rise to regular behaviour. Laws represent this regular behaviour of nomological machines, which implies that those laws hold as a consequence of the repeated, successful operation of nomological machines. Therefore, laws – understood as a necessary regular association between properties – do not necessarily hold for the world beyond the nomological machine, which also means that laws do not necessarily exist as independent entities in nature. In the vocabulary of this chapter, the physical phenomenon produced by the technological instrument may not exist independent of (crucial aspects of) the instrument.

On the basis of her analysis of the role of nomological machines in science, Cartwright rejects the view that is held by many philosophers of science, that laws are basic to our scientific knowledge, and that other things happen on account of them. In Cartwright's view, *capacities* are basic, and things happen on the account of capacities that are exerted in particular physical circumstances. Laws arise following the repeated operation of a system of components with stable capacities in particular fortunate circumstances. Therefore, according to Cartwright, our most wide-ranging scientific knowledge is not knowledge of laws but knowledge of the natures of things, which includes knowledge that allows us to build new nomological machines, or, in my vocabulary: technological instruments that produce a specific physical phenomenon such as elastic or superconductive behaviour.

The views presented so far, allow for a shift to the second philosophical perspective (developed in Sect. 7), which assumes that scientific research also aims at the production and theoretical understanding both of physical phenomena and of the instruments creating them. Hacking's (1992) taxonomy of experiments proposes that scientific practice produces (1) instruments, (2) theoretical knowledge that interprets the working of these instruments and explains how these instruments bring about a phenomenon of interest, and (3) theoretical knowledge of how data produced by these instruments need to be interpreted and processed. Yet, Hacking's account does not explain how scientific knowledge can travel beyond the coherent structure in which it has emerged. Cartwright's account, on the other hand, indicates that we acquire knowledge of the physical and instrumental conditions at which certain capacities will exert themselves, which explains how we are able to theoretically predict and interpret the working of instruments and phenomena at new circumstances.

6.4 Other Types of Theory-Ladeness of Instruments

Radder (2003a) has edited a collection that explicitly focuses on the role of technological instruments in scientific experiments. One of its topics is the characterization of different types of theory-ladeness of observation by means of instruments (e.g. Baird 2003; Heidelberger 2003; Hon 2003; Lange 2003, Radder 2003b; Rothbart 2003). Heidelberger agrees with Kuhn that any observation is coloured through a paradigm, that is, through the cognitive background that is required to make an observation. Nonetheless, Baird, Hon, Heidelberger, Lange and Rothbart agree on the idea that a distinction can be made between the theoretical understanding of instruments that produce, construct or imitate phenomena, which is an understanding at the causal, phenomenological or instrumental level, and the theoretical interpretation of the observations made by means of these instruments.

However, the idea that a distinction is possible between the theoretical understanding of instruments and the theoretical interpretation of observations made by means of these instruments, may be less straightforward than these authors suggest. Their view involves a commonly accepted idea, which is well expressed in Radder's definition of experiment: "An experimenter tries to realize an interaction between an object and some apparatus in such a way that a stable correlation between some features of the object and some features of the apparatus will be produced" (Radder 2003b, 153). This definition assumes a clear distinction between the object and the apparatus: But what about the role of the *interaction* between the two? Does not this interaction *produce* a physical phenomenon of which the observed or measured data are manifestations (also see Bogen and Woodward 1988, and Woodward 2003, 2011)? In other words, how do we know that the manifested phenomenon is a characteristic of the object, rather than a phenomenon that only results from the interaction between object and apparatus? As a consequence, is it really possible to clearly distinguish between the theoretical understanding of an instrument and the observations made of the object by means of it?

Similarly, the assumption that *causal understanding* of the object is derived from observations made by means of instruments (e.g., Hacking 1983; Woodward 2003) relies on the idea that observations of the object are produced by means of an interaction between the object and the instrument. Again, what is the contribution of the interaction between the object and the instrument (e.g., between the electron and the apparatus)? How do we know that instruments present us with causal understanding of the object? For, when using very sophisticated instruments in scientific experiments it is not always obvious where 'object', 'Nature' or 'world' begins and technology ends. Can they be distinguished at all? In many cases, rather than manipulating the object under study in our experiments, the apparatus is manipulated. How do we know that – and *how* exactly – the object is manipulated by means of the instrument?

Rom Harré (1998, 2003) argues that the mentioned distinction is only legitimate for situations in which an ontological distinction can be made between the object under study and the instrument used to examine it. In his view, this is possible for observation of properties of objects by means of microscopes and thermometers, since the instrument and the object under study can be clearly distinguished. But it is more difficult with observations made by means of a cyclotron. Below (in Sect. 7.1), his view will be outlined in a bit more detail.

The issue of whether a clear distinction between instrument and object is always possible also leads to the question of whether theoretical knowledge can be generated about the object independent of the instruments by means of which the object has been studied. The problems raised suggest that the meaning of a theoretical concept may be local and entangled with a description of the technological instrument. Differently put, we may ask whether theoretical concepts produced in a specific experimental setting have a 'non-local' meaning. Radder (1996, 2003b) argues against this suggestion. He defends the view that theoretical concepts can be abstracted from the original experimental practice since their meaning becomes non-local as soon as an experimental result is replicated in completely different circumstances – therefore, according to Radder, the theoretical meaning cannot be reduced to the technological level. Below, I will sketch arguments in favour of the opposite view, which assumes that theoretical concepts acquire part of their meaning from the operational definitions of the instruments (also see Chang 2009a). What is more, the empirical and theoretical understanding of the instruments as part of the meaning of theoretical concepts actually explains the possibility of 'replicating experimental results under completely different circumstances.'

7 A Phenomenon-Oriented Perspective: The Material Role of Instruments in Science

7.1 Epistemic Functions of Instruments

In a theory-oriented positivistic view, only two elements of science are considered: observations and theories. Instruments and experiments are used to test theories, theories are about 'nature' or 'world', and instruments can be clearly distinguished from the object ('Nature' or 'world') under study. So far, the theory-ladeness of observations by means of instruments has been discussed. In Sect. 6.4, a related but different question emerged, namely: To *what* does the empirical and theoretical knowledge produced in modern physical experiments relate? Does it relate to the object under study, or to the phenomenon produced through the interaction between the object under study and the instrument? Hence, the alleged distinction between the instrument and the object, property or process under study is problematic for several reasons: (1) usually, instruments cannot be considered as a mere window on the world; (2) often there is an interaction between the object, property or process under study and the instruments by means of which it is investigated; and (3) in some cases, this interaction, or even the mere technological instrument, produces a phenomenon that does not exist in nature, but becomes itself the object of investigation.

Apparently, instruments have different kinds of epistemic roles in experimental practices. This makes relevant a classification of technological instruments according to their epistemic function. Heidelberger (2003) distinguishes two basic forms of experiments. Firstly, in a theoretical context, instruments have a representative role; their epistemic goal is to *represent* symbolically the relations between natural phenomena and thus to better understand how phenomena are ordered and related to each other. Examples of these instruments are clocks, balances, and measuring rods. Secondly, in causal manipulation by means of instruments, these instruments are used in discoveries and can be distinguished between (a) instruments that have a constructive function (when phenomena are manipulated) and imitative instruments (producing effects in the same way as they appear in nature, without human intervention), and (b) instruments that are used to fulfil a productive function of phenomena that are usually not in the human experience; these are either known phenomena although in circumstances where they have not appeared before – e.g. microscopes –, or unknown phenomena – e.g. Roentgen's production of unknown effects.

In order to address cases in which objects and instruments are entangled, Harré (2003) proposes a classification of instruments based on distinct ontological relationships between laboratory equipment and 'the world'. In his view epistemic functions are derived from ontological relationships. Firstly he distinguishes between *instrument* and *apparatus*. An 'instrument' is defined as detached from the world to be studied, whereas an 'apparatus' is defined as being part of it. (1) 'Instruments' measure either primary qualities (e.g. representing a shape with a microscope) or secondary qualities (e.g. measuring temperature with a thermometer or detecting the presence of acidity with litmus paper). (2) 'Apparatus', on the other hand, are material models of the systems of the world. Regarding 'apparatus', Harré distinguishes between (a) those that are domesticated versions of natural systems (e.g. Drosophila colonies in the laboratory), and (b) Bohrian artefacts that produce phenomena which should not be regarded as the manifestation of a potentiality in the world but as properties of a novel kind of entity, the 'apparatus/world complex'.

Based on the similarities of classifications by authors such as Heidelberger and Harré, I propose a distinction between three types of technological instruments in scientific practices, which I have called *Measure*, *Model* and *Manufacture*

(Boon 2004). 'Measure' is a category of instruments that measure, represent or detect certain features or parameters of an object, process or natural state. 'Model' is a type of laboratory system designed to function as a material model of either natural or technological objects, processes or systems. 'Manufacture' is a type of apparatus that produces a phenomenon that is either conjectured from a new theory or a newly produced phenomenon not as yet theoretically understood. The distinction between these types is based on differences in their epistemic and material function in experimental research. 'Measure's function is to generate data, for instance values of physical variables under specified conditions. 'Model's function is to generate scientific knowledge about a model system, either natural or technological. Somewhat provocatively, I suggested in Boon (2004) that instruments of the manufacture type aim at ontological claims, that is, experiments with manufacture type of devices aim to demonstrate the existence of building blocks or fundamental processes in physical reality. An example of the latter is super-conduction.

Currently, I would add that this interpretation of the epistemic role of instruments of the manufacture type is too narrow. The given interpretation supports a clear distinction between the object and the instrument, and as such it is appropriate to assume that the instrument enables the discovery of the object. But at present, I deny that it makes sense to think of a phenomenon such as super-conduction as existing independent of crucial aspects of the technological instruments producing it. Many of the physical phenomena studied in our laboratories require very specific physical conditions to manifest – these conditions may either occur in nature, or be reproduced by means of technological instruments, or even, never occur in nature but only by means of technological instruments. In any case, as I will argue below, such phenomena are always understood in terms of the physical conditions that bring them about.

My current view on the epistemic role of instruments is close to Cartwright's notion of nomological machines and Harré's ideas of Bohrian artefacts. Yet, recognizing that many of the phenomena discovered and investigated in our laboratory are of technological interest asks for a broader perspective on their role in scientific practices. They are no longer just a means for proving theories. Instead, in technoscientific research, the development of theories is often entangled with the development of instruments that produce technologically relevant phenomena.

7.2 Epistemic Things and Tools

Although Hacking emphasizes both intervention with and the materiality of instruments and studied objects, and Cartwright (1983, 1989, 1999) stresses that scientific laws are interconnected with the instruments that produce them, their views are still primarily embedded in the 'theory-oriented' perspective on science. Rheinberger (1997) broadens the perspective. Similar to Hacking and Cartwright, he emphasizes materiality and manipulability as well as the interconnectedness of instruments and knowledge – but he also enlarges the epistemic role of technological instruments:

"Experimental systems are to be seen as the smallest integral working units of research. As such they are systems of manipulation designed to give unknown answers to questions that the experimenters themselves are not yet able clearly to ask. ... They are not simply experimental devices that generate answers; experimental systems are vehicles for materializing questions. *They inextricably cogenerate the phenomena or material entities and the concepts they come to embody.* Practices and concepts thus 'come packaged together.' ... It is only the process of making one's way through a complex experimental landscape that scientifically meaningful simple things get delineated; in a non-Cartesian epistemology, they are not given from the beginning" (Rheinberger 1997, 28, my emphasis).

For describing this epistemic function of experiments and technological instruments, Rheinberger introduces the notion of 'epistemic things' (also see Baird and Thomas 1990).

Here it is important to distinguish Rheinberger's (1997) notion of 'epistemic things' from my own notion of 'epistemic tools' (Boon 2012, and see below). Whereas an epistemic tool is a tool for thinking (e.g. for thinking about possible interventions with phenomena and/or instruments, and for predicting the outcomes of those interventions), an epistemic thing in Rheinberger's writing is a technological thing, an experimental device, a research object or a scientific object. Epistemic things are material entities or processes – physical structures, chemical reactions, biological functions – that constitute the objects of inquiry. Epistemic tools, on the other hand, are descriptions or pictures (e.g., empirical knowledge, laws, theoretical concepts, scientific models and theories) that are constructed such that we can use them in performing epistemic tasks.

Although the use of these two terms may be somewhat confusing, they add to each other in an account of scientific practices that explains, firstly, in what sense the development of technological instruments and knowledge go hand-in-hand, and secondly, how it is possible that the results of scientific research can be utilized and further developed into high-tech applications.

7.3 Theoretical Concepts and Technological Instruments

In the introduction to this chapter, I suggested that alongside their epistemological function in developing and testing scientific theories and models, technological instruments play a material role in producing specific physical phenomena that may be of interest for developing new technological functions. In a phenomenon-oriented perspective on science, scientific research aims at measuring and/or producing physical phenomena by means of technological instruments, and at understanding both the phenomena and the workings of the instruments to such extent that they (i.e., the instrument and the phenomenon) can be built, controlled, created, calibrated and/or otherwise manipulated.

In a theory-oriented perspective on science, the theory-ladenness of experimental observations in testing theories is problematic. Sects. 4, 5, and 6 presented an outline of how authors in the *New Experimentalist* movement include the role of instruments and experiments when addressing this problem. Yet, accounting for the role of instruments makes the problem worse, as it has become very obvious that instruments are not windows onto the world but are productive themselves. Chang (2004) even argues that the role of instruments in measurement is a locus where "the problems of foundationalism are revealed with stark clarity." The chance to solve the theory-ladeness problem is thus greatly reduced. In a 'phenomenon-oriented' perspective, attention is paid to the role of scientific instruments in the material production of physical phenomena. The theory-ladeness of observing these phenomena in experiments may thus be less problematic. Nonetheless, even in this alternative perspective the question remains of how we should account for *observing new phenomena*, as 'observing' them involves the formation of a theoretical concept, which requires theory. In other words, how are new, (technologically relevant) *theoretical concepts* such pseudo-elasticity and super-conductivity formed?

Hacking (1992) argues that observations, instruments and theories are 'tailored together' in order to produce a stable fit, but does not explain how the tailoring together occurs. Chang (2004) convincingly shows by means of detailed historical studies that the development of an instrument (the thermometer) is entangled with the production of empirical and theoretical knowledge (in thermodynamics). As an alternative to what he calls 'foundationalism', Chang proposes "a brand of coherentism buttressed by the method of 'epistemic iteration'. In epistemic iteration we start by adopting an existing system of knowledge, with some respect for it but without any firm assurance that it is correct; on the basis of that initially affirmed system we launch inquiries that result in the refinement and even correction of the original system. It is a self-correcting progress that justifies (retrospectively) successful courses of development in science, not any assurance by reference to some indubitable foundation." (Chang 2004, 6)

In a similar way, several authors have drawn close connections between experimentation and the formation of theoretical concepts. In this manner, they aim to explain: (1) how theoretical concepts are formed in experimental practices, (2) how the formation of scientific concepts goes hand-in-hand with the development of technological instruments (e.g., Feest 2008, 2010, 2011), (3) how theoretical concepts themselves play a role in investigating the phenomena to which they supposedly refer (e.g., Feest 2008, 2010, and Boon 2012), and conversely (4) how material objects are the driving forces in the process of knowledge acquisition (e.g., Rheinberger 1997; Chang 2009b), and also, (5) how in these processes phenomena and theoretical concepts get stabilized (Chang 2009b; Feest 2011).

Feest (2010) makes an important contribution by considering a theoretical concept not firstly as a *definition* of the purported object of research, but as a *tool* in the process of investigating it. In her opinion, theoretical concepts are operational definitions (also see Chang 2009a), which function as tools to this end by providing the paradigmatic conditions of application for the concepts in question. These are cast in terms of a description of a typical experimental set-up thought to produce data that are indicative of the phenomenon picked out by the concept. Accordingly, theoretical concepts are formulated in terms of supposed crucial aspects of the experimental set-up, which includes the technological instruments, and in that manner, according to Feest, they are tools which allow for experimental interventions into the domain of study, thereby generating knowledge about the phenomenon. Like other tools they can be adapted or discarded in the process.

Closely related to this line of thought, Chang (2009b) asks why some epistemic objects (cf. Rheinberger 1997) persist despite undergoing serious changes, while others become extinct in similar situations? Based on historical studies, he defends the idea that epistemic objects such as 'oxygen' have been retained due to a sufficient continuity of meaning to warrant the preservation of the same term (in spite of major changes of its meaning), but only at the operational level and not at the theoretical level. Furthermore, Chang argues that it might have done some good to keep phlogiston beyond the time of its actual death. Although Chang does not use this vocabulary explicitly, the reason is that 'phlogiston' could have functioned as an epistemic tool enabling different kinds of research questions to those prompted by 'oxygen'. Amongst other things, phlogiston could have served "as an expression of chemical potential energy, which the weight-obsessed oxygen theory completely lost sight of."[5]

In a similar line, Joe Rouse (2011), asks the old question of how theoretical concepts acquire content from their relation to experience – how does 'conceptual articulation' occur? Rouse's critical remark about a positivistic philosophy of science is that "before we can ask about the empirical justification of a claim, we must understand *what* it claims." According to Rouse, positivistic philosophy of science treats conceptual articulation as an entirely linguistic or mathematical activity of developing and regulating inferential relations among sentences or equations. In opposition, Rouse argues that conceptual articulation and empirical justification by means of experimentation and observation cannot be divided in this manner. He first argues that the problem of observation should be transformed by understanding the sciences' accountability to the world in terms of experimental and fieldwork practices. Secondly, he points out that this transformation shows that conceptual articulation is not merely a matter of spontaneous thought in language or mathematics (and thus not merely intra-linguistic); instead, experimental practice itself can contribute to the articulation of conceptual understanding (see for similar ideas, Van Fraassen 2008, 2012, and Massimi 2011).

---

[5] Also see Pickering (1984), who makes a claim similar to Chang (2009a, b). Based on his historical analysis of how the concept of quark has been constructed, he argues that the emergence of the quark idea was not inevitable. He believes that in the early 1970s some options were open to high-energy physics such that physics could have developed in a non-quarky way. Important to this argument is his denial of the view of experiments as the supreme arbiter capable of proving scientific claims, for instance, to the existence of entities such as quarks. Instead, according to Pickering, the quark idea established within a preferred theoretical framework (the gauge theory) that affects how a possible interpretation of experimental data is judged.

Also see Boon (forthcoming), in which I analyze the controversy between Pickering (1984) and Hacking (2000) on the question of to which extent the entities in successful theories are inevitable or contingent. As an alternative to Hacking's realism, I propose epistemological constructivism, which is in alignment with those of Chang (2009b) and Rouse (2011).

Following up on these ideas, I have proposed that the formation of theoretical concepts involves interplay between experimental observations and 'partial' conceptual, empirical and theoretical knowledge of the working of an instrument or experimental set-up. At the outset, the latter kinds of knowledge enable scientists to recognize the experimental observations as of a certain type of phenomenon, such as elastic behaviour or electrical conduction of a material. The relevant point is that the initial interpretation of the data is enabled and guided by conceptual, empirical and theoretical knowledge of the instrument and/or experimental set-up. At the same time, this knowledge enables the scientist to recognize that the experimental observations are at variance with empirically known behaviour, thus pointing at a new kind of physical phenomenon. Finally, formation of a theoretical concept of a phenomenon involves interpreting the experimental observations by employing relevant conceptual, empirical and theoretical background knowledge. In other words, the leap from experimental observations to new theoretical concepts for describing new kinds of physical phenomena builds on empirical knowledge and theoretical understanding of the instruments and experimental set-up. As a consequence, theoretical concepts remain connected with knowledge of supposedly relevant physical conditions and aspects of the instrument and/or experimental set-up by means of which it has been produced.

8 Conclusions

In this chapter, my goal is not an exhaustive overview of literature in the philosophy of science that addresses the role of technological instruments in science. Rather, I have aimed to present an overview of ideas that explain the entangled epistemic and material role of instruments and experiments in scientific research practices. For a long time, the philosophy of science has ignored the role of technological instruments. What is more, both scientists and philosophers have left us with a sense of awe but also confusion about the technological achievements of science. How is it possible that, by means of mere theoretical knowledge we can design and build previously inconceivable technological devices? These apparent miraculous achievements have been an important argument for the so-called miracle argument in the philosophy of science: such incredible successes can only be explained when assuming that our most successful scientific theories are true and the entities proposed in these theories really exist. Yet, based on a better understanding of the role of technological instruments in the formation of scientific knowledge (such as, theoretical concepts, empirical knowledge and scientific models of phenomena), alternative explanations come into view. Possibly, an explanation of the remarkable technological achievements could be less reliant on the character of theories, and more dependent on the crucial material and epistemic role of technological instruments in scientific research practices.

Acknowledgments The writing of this article was financially supported by a Vidi grant from the Dutch National Science Foundation NWO. I am grateful to the ZIF Bielefeld and the interdisciplinary research group "Science in the Context of Application" who supported part of this research. I would also like to thank my language editor, Claire Neesham at Centipede Productions.

References

Ackermann, R. J. (1985). *Data, instruments and theory: A dialectical approach to understanding science*. Princeton: Princeton University Press.
Ackermann, R. (1989). The new experimentalism. *The British Journal for the Philosophy of Science, 40*(2), 185–190.
Baird, D. (2003). Thing knowledge: Outline of a materialist theory of knowledge. In H. Radder (Ed.), *The philosophy of scientific experimentation* (pp. 39–67). Pittsburgh: University of Pittsburgh Press.
Baird, D. (2004). *Thing knowledge: A philosophy of scientific instruments*. Berkeley: University of California Press.
Baird, D., & Thomas, F. (1990). Scientific instruments, scientific progress and the cyclotron. *British Journal for the Philosophy of Science, 41*(2), 147–175.
Bogen, J., & Woodward, J. (1988). Saving the phenomena. *The Philosophical Review, 97*(2), 303–352.
Boon, M. (2004). Technological instruments in scientific experimentation. *International Studies in the Philosophy of Science, 18*(2&3), 221–230.
Boon, M. (2011). In defence of engineering sciences: On the epistemological relations between science and technology. *Techné: Research in Philosophy and Technology, 15*(1), 49–71.
Boon, M. (2012). Scientific concepts in the engineering sciences: Epistemic tools for creating and intervening with phenomena. In U. Feest & F. Steinle (Eds.), *Scientific concepts and investigative practice* (Berlin studies in knowledge research, pp. 219–243). Berlin/New York: Walter De Gruyter GMBH & CO. KG.
Boon, M. (forthcoming). Contingency and inevitability in science – Instruments, interfaces and the independent world. In L. Soler, E. Trizio & A. Pickering (Eds.), *Science as it could have been: Discussing the contingent/inevitable aspects of scientific practices*. Pittsburgh: University of Pittsburgh Press.
Carrier, M. (1998). New experimentalism and the changing significance of experiments: On the shortcomings of an equipment-centered guide to history. In M. Heidelberger & F. Steinle (Eds.), *Experimental essays – Versuche zum experiment* (pp. 175–191). Baden-Baden: Nomos.
Carrier, M. (2001). Error and the growth of experimental knowledge. *International Studies in the Philosophy of Science, 15*(1), 93–98.
Cartwright, N. (1983). *How the laws of physics lie*. Oxford: Clarendon Press/Oxford University Press.
Cartwright, N. (1989). *Natures capacities and their measurement*. Oxford: Clarendon Press/Oxford University Press.
Cartwright, N. (1999). *The dappled world. A study of the boundaries of science*. Cambridge: Cambridge University Press.
Chalmers, A. (2003). The theory-dependence of the use of instruments in science. *Philosophy of Science, 70*(3), 493–509.
Chang, H. (2004). *Inventing temperature: Measurement and scientific progress*. Oxford: Oxford University Press.
Chang, H. (2009a). Operationalism. In E. N. Zalta (Ed.), *The Stanford encyclopaedia of philosophy*. http://plato.stanford.edu/archives/fall2009/entries/operationalism/
Chang, H. (2009b). The persistence of epistemic objects through scientific change. *Erkenntnis, 75*, 413–429.

Feest, U. (2008). Concepts as tools in the experimental generation of knowledge in psychology. In U. Feest, G. Hon, H. J. Rheinberger, J. Schickore & F. Steinle (Eds.), *Generating experimental knowledge* (Vol. 340, pp. 19–26). MPI-Preprint.

Feest, U. (2010). Concepts as tools in the experimental generation of knowledge in cognitive neuropsychology. *Spontaneous Generations: A Journal for the History and Philosophy of Science, 4*(1), 173–190.

Feest, U. (2011). What exactly is stabilized when phenomena are stabilized? *Synthese: An International Journal for Epistemology, Methodology and Philosophy of Science, 182*(1), 57–71.

Frigerio, A., Giordani, A., & Mari, L. (2010). Outline of a general model of measurement. *Synthese: An International Journal for Epistemology, Methodology and Philosophy of Science., 175*, 123–149.

Franklin, A. (1986). *The neglect of experiment*. Cambridge/New York: Cambridge University Press.

Franklin, A. (2002). Experiment in physics. In E. N. Zalta (Ed.), *The Stanford encyclopaedia of philosophy*. http://plato.stanford.edu/archives/win2012/entries/physics-experiment/

Franklin, L. (2005). Exploratory experiments. *Philosophy of Science, 72*(5), 888–899.

Galison, P. (1987). *How experiments end*. Chicago/London: University of Chicago Press.

Galison, P. (1997). *Image and logic: A material culture of microphysics*. Chicago: Chicago University Press.

Hacking, I. (1983). *Representing and intervening: Introductory topics in the philosophy of natural science*. Cambridge: Cambridge University Press.

Hacking, I. (1992). The self-vindication of the laboratory sciences. In A. Pickering (Ed.), *Science as practice and culture* (pp. 29–64). Chicago: University of Chicago Press.

Hacking, I. (2000). How inevitable are the results of successful science? *Philosophy of Science, 67*, S58–S71.

Harré, R. (1998). Recovering the experiment. *Philosophy, 73*(285), 353–377.

Harré, R. (2003). The materiality of instruments in a metaphysics for experiments. In H. Radder (Ed.), *The philosophy of scientific experimentation* (pp. 19–38). Pittsburgh: University of Pittsburgh Press.

Heidelberger, M. (2003). Theory-ladenness and scientific instruments. In H. Radder (Ed.), *The philosophy of scientific experimentation* (pp. 138–151). Pittsburgh: University of Pittsburgh Press.

Hon, G. (1998). Exploiting errors: Review of Deborah Mayo's error and the growth of experimental knowledge. *Studies in the History and Philosophy of Science, 29*(3), 465–479.

Hon, G. (2003). The idols of experiment: Transcending the 'Etc. List'. In H. Radder (Ed.), *The philosophy of scientific experimentation* (pp. 174–197). Pittsburgh: University of Pittsburgh Press.

Ihde, D. (1991). *Instrumental realism: The interface between philosophy of science and philosophy of technology*. Bloomington/Indianapolis: Indiana University Press.

Ihde, D. (2009). From da Vinci to CAD and beyond. *Synthese: An International Journal for Epistemology, Methodology and Philosophy of Science, 168*(3), 453–467.

Ihde, D., & Selinger, E. (Eds.). (2003). *Chasing technoscience. Matrix for materiality*. Bloomington/Indianapolis: Indiana University Press.

Krantz, D. H., Luce, R. D., Suppes, P., & Tversky, A. (1971). *Foundations of measurement* (Vol. 1). New York: Academic.

Kuhn, T. S. (1970). *The structure of scientific revolutions* (2nd ed.). Chicago: The University of Chicago Press.

Lacey, H. (2012). Reflections on science and technoscience. *Scientiae Studia: Revista Latino-Americana De Filosofia E História Da Ciência, 10*, 103–128.

Lange, R. (2003). Technology as basis and object of experimental practices. In H. Radder (Ed.), *The philosophy of scientific experimentation* (pp. 119–137). Pittsburgh: University of Pittsburgh Press.

Luce, R. D., Krantz, D. H., Suppes, P., & Tversky, A. (1990). *Foundations of measurement* (Vol. 3). San Diego: Academic.
Massimi, M. (2011). From data to phenomena: A Kantian stance. *Synthese: An International Journal for Epistemology, Methodology and Philosophy of Science, 182*(1), 101–116.
Mayo, D. G. (1994). The new experimentalism, topical hypotheses, and learning from error. *PSA: Proceedings of the Biennial Meeting of the Philosophy of Science Association, 1994*, 270–279.
Mayo, D. G. (1996). *Error and the growth of experimental knowledge*. Chicago: University of Chicago Press.
Mayo, D. G. (2000). Experimental practice and an error statistical account of evidence. *Philosophy of Science, 67*(3), S193–S207.
Nordmann, A. (2012). Object lessons: Towards an epistemology of technoscience. *Scientiae Studia: Revista Latino-Americana De Filosofia E História Da Ciência, 10*, 11–31.
Pickering, A. (1984). *Constructing quarks, a sociological history of particle physics*. Chicago/London: University of Chicago Press.
Pitt, J. C. (2000). *Thinking about technology: Thinking about the foundations of a philosophy of technology*. New York: Seven Bridges Press.
Popper, K. R. (1959). *The logic of scientific discovery*. London: Hutchinson.
Radder, H. (1996). *In and about the world: Philosophical studies of science and technology*. Albany: State University of New York Press.
Radder, H. (2003a). Toward a more developed philosophy of scientific experimentation. In H. Radder (Ed.), *The philosophy of scientific experimentation* (pp. 1–18). Pittsburgh: University of Pittsburgh Press.
Radder, H. (2003b). Technology and theory in experimental science. In H. Radder (Ed.), *The philosophy of scientific experimentation* (pp. 152–173). Pittsburgh: University of Pittsburgh Press.
Radder, H. (2004). Book review: Chasing technoscience. Matrix for materiality. *Philosophy of Science, 71*(4), 614–619.
Rheinberger, H. J. (1997). *Toward a history of epistemic things: Synthesizing proteins in the test tube*. Stanford: Stanford University Press.
Rothbart, D. (2003). Designing instruments and the design of nature. In H. Radder (Ed.), *The philosophy of scientific experimentation* (pp. 236–254). Pittsburgh: University of Pittsburgh Press.
Rouse, J. (1987). *Knowledge and power: Towards a political philosophy of science*. Ithaca: Cornell University Press.
Rouse, J. (2011). Articulating the world: Experimental systems and conceptual understanding. *International Studies in the Philosophy of Science, 25*(3), 243–254.
Steinle, F. (2002). Challenging established concepts: Ampere and exploratory experimentation. *Theoria, 17*(44), 291–316.
Steinle, F. (2006). Concept formation and the limits of justification: The two electricities. *Archimedes, 14*(III), 183–195.
Suppes, P., Krantz, D. H., Luce, R. D., & Tversky, A. (1990). *Foundations of measurement* (Vol. 2). New York: Academic.
Van Fraassen, B. (2008). *Scientific representation*. Oxford: Oxford University Press.
Van Fraassen, B. (2012). Modeling and measurement: The criterion of empirical grounding. *Philosophy of Science, 79*(5), 773–784.
Woodward, J. F. (2003). Experimentation, causal inference, and instrumental realism. In H. Radder (Ed.), *The philosophy of scientific experimentation* (pp. 87–118). Pittsburgh: University of Pittsburgh Press.
Woodward, J. F. (2011). Data and phenomena: A restatement and defense. *Synthese, 182*(1), 165–179.

Chapter 5
Experiments Before Science. What Science Learned from Technological Experiments

Sven Ove Hansson

Abstract Systematic experimentation is usually conceived as a practice that began with science, but this assumption does not seem to be correct. Historical evidence gives us strong reasons to believe that the first experiments were not scientific, but instead directly action-guiding technological experiments. Such experiments still have a major role for instance in technology and agriculture and (in the form of clinical trials) in medicine. The historical background of such experiments is tracked down, and their philosophical significance is discussed. Directly action-guiding experiments have a strong and immediate justification and are much less theory-dependent than other (scientific) experiments. However, the safeguards needed to avoid mistakes in the execution and interpretation of experiments are essentially the same for the two types of experiments. Several of these safeguards are parts of the heritage from technological experiments that science has taken over.

1 Introduction

It is generally recognized that scientific experiments rely on technology in at least two ways. First, experiments make use of increasingly complex technological equipment. This has been a major argument for describing science as applied technology, rather than the other way around (Lelas 1993). Secondly, the development of experimental methodology in Renaissance science depended to a large extent on contributions by skilled craftsmen. Galileo Galilei (1564–1642) was in close contact with craftsmen and learned from their experience, and quite a few craftsmen made important contributions of their own to the early development of natural science (Zilsel 1941, 1942, 2000). Francis Bacon (1561–1626), perhaps the most influential philosophical spokesperson of the emerging experimental science, defended the often disdained knowledge of craftspeople and emphasized its usefulness in scientific inquiry (Bacon [1605] 1869, 88–89). To this I will add a third connection. I will show that experiments began in technology. The first experiments

---

S.O. Hansson (✉)
Division of Philosophy, Royal Institute of Technology (KTH), Brinellvägen 32, 10044 Stockholm, Sweden
e-mail: soh@kth.se

were directly action-guiding experiments that were developed for technological purposes. Moreover, the justification of experimental procedures is much stronger for these experiments than for the experiments undertaken in order to gain better understanding of the workings of nature. Therefore, in order to understand scientific experimentation we must pay close attention to its origins in technology.

2 What Is an Experiment?

To begin with, what do we mean by an experiment? There is an everyday usage of the term that has very little to do with science or technology, for instance when we say "He is experimenting with drugs". This is not a modern extension of the meaning. To the contrary, the original meaning of "experiment" was quite wide. The word has the same origin as "experience", namely the latin verb "experiri" that means to try or put to test.

In his *Tusculan Disputations* Cicero (106–43 BCE) used the fact that grief is removed by the passage of time as an argument to show that there is no real evil in that which one grieves. "Hoc maximum est experimentum" he said, "this is the highest proof" (III:XXX:74). Today we would call this a proof by experience rather than by experiment. And this is indeed how the word was commonly used. In the original sense of the word, an experiment could be just a passive observation, not requiring the investigator to have any active role.

Roger Grossetest (1175–1253) introduced the term "experiment" into the discussion of how we can attain knowledge in empirical matters, i.e. what we now call scientific method. However, he did not reserve the term for active procedures but included the passive collection of experiences (Eastwood 1968, 321; McEvoy 1982, 207–208). This broad sense of "experiment" can be followed through the history of philosophical discussions of empirical knowledge. In his *Novum Organum* Francis Bacon (1561–1626) likewise emphasized the testing of claims about nature against observations, but he used the word "experiment" for passive observations with instruments as well as for active procedures (Klein 1996, 293; Pesic 1999; Klein 2005).

David Hume's (1711–1776) *Treatise of Human Nature* had the subtitle "Being an Attempt to Introduce the Experimental Method of Reasoning into Moral Subjects". However some of his experiments take the form of recording everyday experiences, and would not be counted as experiments in the modern sense (Robison 2008).

The English astronomer and polymath John Herschel (1792–1871) drew the distinction between observation and experiment very clearly in his influential *A preliminary discourse on the study of natural philosophy* from 1831:

> We have thus pointed out to us, as the great, and indeed only ultimate source of our knowledge of nature and its laws, EXPERIENCE... But experience may be acquired in two ways: either, first, by noticing facts as they occur, without any attempt to influence the frequency of their occurrence, or to vary the circumstances under which they occur; this is OBSERVATION: or, secondly, by putting in action causes and agents over which we have

control, and purposely varying their combinations, and noticing what effects take place; this is EXPERIMENT. (Herschel 1831, 76)

In his *System of Logic* (1843) John Stuart Mill made a distinction between experiment and observation, and endorsed experiment as the most useful method.

> The first and most obvious distinction between Observation and Experiment is, that the latter is an immense extension of the former. It not only enables us to produce a much greater number of variations in the circumstances than nature spontaneously offers, but also, in thousands of cases, to produce the precise sort of variation which we are in want of for discovering the law of the phenomenon; a service which nature, being constructed on a quite different scheme from that of facilitating our studies, is seldom so friendly as to bestow upon us. For example, in order to ascertain what principle in the atmosphere enables it to sustain life, the variation we require is that a living animal should be immersed in each component element of the atmosphere separately. But nature does not supply either oxygen or azote [nitrogen] in a separate state. We are indebted to artificial experiment for our knowledge that it is the former, and not the latter, which supports respiration; and for our knowledge of the very existence of the two ingredients. (Mill [1843] 1974, 382.)

With this we have arrived essentially at the modern notion of an experiment. It was expressed very clearly by William Stanley Jevons (1835–1882) in his *Principles of Science* from 1874:

> We are said to *experiment* when we bring substances together under various conditions of temperature, pressure, electric disturbance, chemical action, &c., and then record the changes observed. Our object in inductive investigation is to ascertain exactly the group of circumstances or conditions which being present, a certain other group of phenomena will follow. If we denote by A the antecedent group, and by X subsequent phenomena, our object will usually be to discover a law of the form $A = AX$, the meaning of which is that where A is X will happen. (Jevons 1920, 416)

As already mentioned, this is the modern meaning of "experiment". Admittedly, the older, wider meaning that includes passive observation has not disappeared. Well into the 1930s many writers of academic textbooks used the term "experimental" as synonymous with "empirical" (Winston and Blais 1996). Examples of the usage can also be found in the modern scholarly literature.[1] But since I need the distinction, in what follows I will use the term in its modern, more restricted sense. In other words, by an experiment I will mean a procedure in which some object of study is subjected to interventions (manipulations) that aim at obtaining a predictable outcome or at least predictable aspects of the outcome. Predictability of the outcome, usually expressed as repeatability of the experiment, is an essential component of the definition. Experiments provide us with information about regularities, and without predictability or repeatability we do not have evidence of anything regular. A procedure that we carry through can only have the intended function of an experiment in our deliberations if its setup is so constructed that it determines at least some aspects of the outcome.

---

[1]For instance in Prioreschi (1994, 145). For a criticism of this usage in current so-called experimental philosophy, see Hansson (2014).

This was very well expressed by Jürgen Habermas in his *Knowledge and Human Interests*:

> In an experiment we bring about, by means of a controlled succession of events, a relation between at least two empirical variables. This relation satisfies two conditions. It can be expressed *grammatically* in the form of a conditional prediction that can be deduced from a general lawlike hypothesis with the aid of initial conditions; at the same time it can be exhibited *factually* in the form of an instrumental action that manipulates the initial conditions such that the success of the operation can be controlled by means of the occurrence of the effect. (Habermas 1968, 162, 1978, 126)

3 Two Types of Experimental Inquiry

Science can be conducted with two major types of goals that we can refer to with the traditional terms techne and episteme. Techne (in this interpretation) refers to goals in practical life, and episteme to knowledge or understanding of the workings of the world. A large part of science aims at knowledge of the former type, knowledge about how we can achieve various practical aims. Medical, agricultural and technological research all aim primarily at finding ways to achieve practical goals outside of science, rather than knowledge goals. I propose that there is a specific type of experiments[2] that we use to obtain knowledge of the techne type. Consider the following examples:

- Ten different types of steel are tested in order to find out which of them is most resistant to fatigue.
- In order to determine the longevity of different types of lightbulbs, a test is performed in which lightbulbs of different types are continuously lit in the same room, and their times of extinction are recorded.
- In an agricultural field trial, two varieties of wheat are sown in adjacent fields in order to find out which of the cultivars is best suited for the local conditions.
- In a clinical trial, alternative treatments are administered to different groups of hypertension patients in order to find out which of the treatments is most efficient in reducing blood pressure.
- In a social experiment, different social work methods are tested in different communities in order to see which of them has the best effect on drug abuse.

These experiments all have two things in common. First, they have *an outcome measure that coincides with some desired goal of human action*. We measure the outcome of the agricultural field trial according to the yield (and other criteria of agricultural success), the outcome of the clinical trial in terms of whether the patients recover from the disease, etc. This is very different from other scientific experiments, in which the outcome measure is chosen so that it can provide us with

---

[2]More precisely: type of interpretations of experiments.

information about the workings of nature. When we measure the wavelength of electromagnetic radiation emitted in an experiment in nuclear physics, this is not done in order to obtain some specific wavelength but to obtain information that we can use to understand the reaction.

The other feature that is common for this type of experiment is that *the experimental setup is intended to realize our preliminary hypothesis about what type of action or intervention would best achieve the desired effect*. We perform field trials with the cultivars that we believe to yield the best harvests, clinical trials with the therapies that we believe to be best for the patients, social experiments with the social work methods that are considered to be most promising, etc.

I will call these experiments *directly action-guiding* because they are constructed to show us directly to what extent a potential line of action will have its desired effects. The other major type of experiments can be called *epistemic* since their aim is to provide us with better knowledge and understanding. Obviously, experiments of the latter type can also guide action, but only in more indirect ways.

Although it is convenient to talk about these as two categories of experiments, they are in fact two categories of intended interpretations of experiments. Therefore, one and the same experiment can belong to both categories. The best examples that I have found of such double use belong to the disciplines of experimental archaeology and experimental history of technology. In order to find out whether a particular stone age tool could have been used for felling trees we had better make replicas and try systematically to fell trees with them. If the same experiment had been performed 5,000 years ago it would presumably have been a directly action-guiding experiment, aimed at guiding the choice of tools. But when performed today, it is an epistemic experiment since it aims at providing us with knowledge about the lives of our stone age ancestors. Similarly, if a historian performs an experiment to compare the efficiency of two types of scythes, then that is an epistemic experiment. If an Amish farmer makes the same test, then it is probably a directly action-guiding experiment.

We can summarize the idea behind directly action-guiding experiments in the form of a simple "recipe":

Recipe for directly action-guiding experiments

If you want to know if you can achieve Y by doing X, do X and see if Y occurs. I will return below to the question what the corresponding recipe for epistemic experiments can be. But before that I will present and argue for three theses about directly action-guiding experiments. The first of these concerns the history of experimentation.

4 The Early History of Experimentation

Epistemic experiments were rare before the Renaissance, and when focusing only on epistemic experiments we find ourselves discussing questions such as whether the ancient Greeks performed experiments. If we include directly action-guiding experiments then we will find that experimentation goes back to prehistoric times. Long before there was science there was experimentation. As I will soon exemplify, farmers and craftspeople performed directly action-guiding experiments in order to find out how they could best achieve their practical aims. Therefore, contrary to what is commonly assumed, the origin of experimentation is neither academic nor curiosity-driven. Instead, it is socially widespread and driven by practical needs. Furthermore, the directly action-guiding experiments of ancient times were technological. Therefore, contrary to what is commonly assumed, technological experiments predated scientific experiments, probably with thousands of years. Let us look at some examples.

4.1 Agricultural Experiments

Agriculture is the crucial technological innovation that made civilization possible. Archaeological studies of ancient agriculture confirm what is biologically obvious, namely that agriculture was not the result of a few inventions or discoveries but of thousands of years with extensive and continuous experimentation. In pre-colonial Latin America, for instance, there was an early period of tropical agriculture without the crops that later became dominant (maize and manioc), characterized by wide variations that are best explained by continuous experimentation. As noted by one prominent archaeologist, the evidence suggests "that people everywhere began by experimenting with the cultivation of plants they were already collecting in the wild" (Bray 2000). Experimentation was necessary for the simple reason that in many climates it was extremely difficult to succeed in growing wild plants and get enough food from them. For instance, the traditional strategy of Andean agriculture has been described as a "massive parallelism" that arose in response to the uncertainties that an unpredictable climate, pests and other natural events give rise to. The most conspicuous part of this strategy was to have as many dispersed fields and as many crops as possible. It was not uncommon for a family to have as many as 30 different fields in different locations and with different microclimatic and ecological conditions. In each of these fields, several different crops were cultivated. This, of course, was a sophisticated risk management strategy; crop failure in some of these fields would be compensated by good harvests in others. As one researcher remarked, "[t]his tactic is probably very ancient and helps to explain the tremendous range of domesticated food crops" (Earls 1998). The obvious way to farm under these conditions must have been to try out different crops in different places and see what worked best. Indeed, that is what Andean farmers traditionally do.

> Agricultural experimentation is virtually universal in the Andes. In every Indian community there are many people who continually experiment with all plants that come their way in special fields (chacras) usually located near their houses, but some do so in a certain chacra in every production zone where they have fields. They watch how different plants 'go' under different climatic conditions. Simulation and experimentation are routine activities for the native Andean agricultor. (Earls 1998)

Close to Cusco, the capital of the Inca empire, on a plateau at 3,500 m altitude, several natural sinkholes have been terraced by the Incas. The site is called Moray. One of the sinkholes is over 70 m deep. The terraces have an irrigation system connected with water supply from the Qechuyoq mountain, but the water was diverted to a neighbouring town in the middle of the twentieth century. According to oral tradition among local peasants, Moray served as "an Inca agricultural college", and the different terrace levels were characterized by "different climates". Recent measurements of soil temperatures confirm this information. There are large climatic differences between different parts of the terrace system, partly due to shadows cast by the Qechuyoq. We have strong reasons to believe that this was indeed a site for agricultural field trials, allowing for the testing of different crops and cultivars under various climatic conditions. What went on here was probably a systematization of traditional experimental practices in Andean agriculture (Earls 1998). We should bear in mind that all this experimentation, among farmers and in the experimental station of Moray, took place in a culture without a written language.

Records from other parts of the world confirm that indigenous and traditional farmers do indeed experiment. This is documented for instance from Mexico (Alcorn and Toledo 1998; Berkes et al. 2000), China (Chandler 1991) and South Sudan (de Schlippe 1956). The Mende people in Sierra Leone have a special word, "hungoo", for experiment. A hungoo can consist in planting two seeds in adjacent rows, and then measuring the output in order to determine which seed was best. This was probably an original practice, not one brought to the Mende by visiting Europeans (Richards 1986, 131–146, 1989). As these examples show, farmers in different parts of the world perform experiments, in traditions that are independent of each other and presumably much older than science.[3]

These experiments correspond to the simple recipe for directly action-guiding experiments that I proposed above. In the farmer's case, the Y of the recipe is usually a good yield of something edible. The X can be sowing or planting in a particular location, sowing at a particular point in time, choosing a particular cultivar, watering more or less, following a particular crop rotation scheme, or any other of the many variables at his or her disposal.

4.2 Craftspeople's Experiments

Just like farmers, craftspeople of different trades have performed directly action-guiding, i.e. technological, experiments since long before modern experimental science. Contrary to scientists, they have not written down their achievements, and

---

[3]For additional references showing that farmers' experiments are indeed widespread, see Johnson (1972). On the need to take farmers' investigative capacities into account in development projects, see Biggs (1980).

therefore very little is known about their experiments. Just as for agricultural experiments, I can only provide some scattered examples. The history of technological experiments remains to be written.

In the early Islamic period, Raqqa (Ar-Raqqah) in eastern Syria was a centre of glass and pottery production. Impressively thorough archaeological investigations have been performed of artefacts and debris found in eighth to eleventh century glassmaking workshops in Raqqa (Henderson et al. 2004, 2005). Several different types of glass were manufactured here. For some of these types, "startling compositional homogeneity" was found "across a number of major and minor constituents" (Henderson et al. 2004, 465). The same batch recipes must have been used for these materials over long periods of times. But for one of the glass types, called by the researchers "type 4", a very different pattern emerged. A wide compositional range was found that corresponds to what would be expected if a large number of experiments were performed in order to improve the quality of type 4 glass. These experiments seem to have included a so-called chemical dilution line in order to determine the appropriate proportions of the main ingredients. They may also have included experiments with additions of other ingredients. The authors conclude that "the compositional variations observed for type 4 glasses are thought to be the result of experimentation first with raw ingredients and then with type 4 cullet" (Henderson et al. 2004, 460).

Judging by the many types of materials and mixtures developed in the course of human history, systematic experimentation to optimize batch recipes could not have been unique to Raqqa but must have taken place in many places throughout the world, in order to determine the best ingredients and proportions for the production of alloys, paints, mortars and a large number of other composite technological materials. As one example of this, the oldest preserved instruction from ancient Egypt on how to make bronze as hard as possible prescribes a mixture of 88% copper and 12% tin. The Egyptians certainly had no theory from which they could deduce this knowledge, so it could only have been found out by trial and error. Since this knowledge was acquired in a rather short period, it was probably obtained through systematic experimentation (Malina 1983). Another example is the composition of mortar. As noted in a recent study of ancient building materials, "[m]asons had to rely on experimentation or information past on orally to understand and learn the properties of the employed materials and the effects they could have produce[d] when added to a mortar" (Moropoulou et al. 2005, 296).

Builders have also experimented with other aspects of their constructions. Let us consider an example from the Roman era. Contrary to what is often believed, a Roman vault such as that of the Pantheon dome exerts a thrust on the columns and the lintels. If the columns and lintels are too slender, then the thrust may cause them to move or even collapse. Various constructions were tried to resist the thrust of the vault, including iron ties that employed the tensile strength of iron to counteract the forces that tend to tear the construction apart. This construction was used for instance in the Basilica Ulpia and the Baths of Caracalla. In both cases, it was combined with the use of lightweight material for the vault. Studies of Roman buildings give strong indications that Roman builders experimented with these and other methods to cope with the thrust from the vaults (DeLaine 1990).

The experimentation of builders did not end with the Romans. Many have wondered how the building of the great Gothic cathedrals beginning in the twelfth century was at all possible. With the construction of new cathedrals, pillars and other construction elements became slimmer and slimmer. How was that possible? No means were available to calculate the required dimensions of the building elements.

Careful analysis of these cathedrals strongly suggests that they were experimental building projects, such that "the design may have been successively modified on the basis of observation of the buildings during the course of construction" (Mark 1972). When a new cathedral was planned, its construction elements were made somewhat slimmer than those of its predecessor. Builders had to be sensitive to signs that the building was not strong enough. One such sign could be cracks in newly set mortar when temporary construction supports were removed. The tensile strength of medieval mortar was extremely low, and therefore the mortar would have revealed also relatively small displacements. Effects of strong winds on the construction were probably also observed. Since the cathedrals took several decades to build, every cathedral must have been subject to severe storms during its construction period. Fortunately, if the construction was found to be too weak, the builders did not have to tear down the cathedral and build a new more sturdy one. Instead they could add construction elements such as flying buttresses that provided the necessary support. In this way, according to one leading analyst "the details of design were worked out with a crude type of experimental stress analysis performed during construction: tensile cracking observed in the weak lime mortar between the stones during the relatively long period of construction could have led to refinements in design" (Mark 1978; cf. Wolfe and Mark 1974).

This exemplifies a note-worthy feature of pre-modern technological experiments: They sometimes lasted much longer than what modern scientific experiments normally do (Snively and Corsiglia 2000).[4] Another impressive evidence of this has been reported from indigenous peasants in the Fujian province on the southeast coast of China. In oral tradition they have kept track of crop rotation in forestry with the China fir (*Cunninghamia lanceolata*). Since the trees have to grow for at least 35 years before they can be harvested, few farmers would survive two rotations of the tree. In spite of this they have accumulated information about soil problems created by repeated planting of these trees three or more times on the same place, and adjusted their silvicultural practices accordingly. The soil problems have been confirmed by modern scientific soil analysis (Chandler 1991). Although this may

---

[4] A few long-term experiments have been performed in modern science. One of the most famous, and possibly the longest running epistemic experiment, is the pitch drop experiment set up in 1927 at the University of Queensland, Australia that is still running. Its purpose is to show that although pitch appears to be a solid it is a high-viscosity fluid. The experiment consists in lettings drops form and fall from a piece of pitch inside a glass container. The eighth drop fell in 2000 and the ninth in 2014. See http://www.smp.uq.edu.au/content/pitch-drop-experiment and Edgeworth et al. (1984).

not be an example of an experiment in the strict sense, it exemplifies the same way of thinking as in the evaluation of a directly action-guiding experiment.

For a final example of ancient technological experimentation, let us return to ancient Peru. Due to the large number of textiles found by archaeologists, it has been possible to perform detailed analyses of technological developments in Peruvian weaving traditions. The introduction of the heddle loom was followed by an innovative period in which many new ways were tried out in order to regain the flexibility of off-loom techniques without losing the labour-saving effects of the heddle. As one researcher noted, "[i]t is highly unlikely that the staggering array of later loom weaves suddenly appeared fully mature with the introduction of the loom. Rather, the new process demanded a long period of patient experimentation, lasting well into the Early Horizon [after 900 BCE.], as weavers tested alternate ways of introducing decorative methods within the heddling system" (Doyon-Bernard 1990).

4.3 Early Medical Experiments

I cannot leave the topic of pre-scientific, directly action-guiding experiments without mentioning medical experiments. The use of herbs and other remedies for diseases can be based on experiments, according to the simple recipe that I mentioned before. If you want to know whether you can reduce a patient's fever by giving her a particular herb, give feverish patients the herb and record if their fever is abated or not. One famous description of a treatment experiment was given by Avicenna (Abd Allah ibn Sina, c. 980–1037):

> [It] is like our judgement that the scammony plant is a purgative for bile; for since this [phenomenon] is repeated many times, one abandons that it is among the things which occur by chance, so the mind judged that it belongs to the character of scammony to purge bile and [the mind] gave into it, that is, purging bile is an intrinsic characteristic belonging to scammony. (McGinnis 2003, 317)

This description of an experiment has the interesting feature of pointing out that the treatment has to be repeated on several patients in order to rule out chance as an explanation.[5] This is important due to the stochastic and variable nature of disease and recovery. But another important feature is missing, and therefore his experiment is arguably not a directly action-guiding experiment. The problem is that the desirability of the outcome measure is far from self-evident. Laxation ("purging bile") is not necessarily desirable. Only if we know that laxation is beneficial for a patient is this experiment action-guiding, and in that case it is, strictly speaking, *in*directly action-guiding.

---

[5]This insight seems to have been missing in another text from the same century. Su Song wrote: "In order to evaluate the efficacy of ginseng, find two people and let one eat ginseng and run, the other run without ginseng. The one that did not eat ginseng will develop shortness of breath earlier." (Claridge and Fabian 2005, 548)

I have not been able to find examples of directly action-guiding experiments on the treatment of diseases before the modern era. Should this surprise us? On one hand, given the widespread use of directly action-guiding experiments in agriculture and the technical crafts we could expect the same methods to have been used in the service of health as well. But on the other hand, the subject-matter of human health tends to be much more imbued with ideology and religion than those of agriculture and technology. This applies to ancient as well as modern societies. Perhaps the down-to-earth nature of questions such as "can I grow this crop here?" or "will the columns support this type of roof?" facilitated the thought processes that led to the construction of directly action-guiding experiments. We may be more prone to look for guidance elsewhere in questions of life or death, health or disease. Possibly, my lack of success in finding examples of directly action-guiding medical experiments among indigenous or ancient peoples may be due to this. Possibly it depends on the limitations either of my search or the historical and anthropological record.

5 The Justification of Directly Action-Guiding Experiments

With this I hope to have substantiated my first thesis about directly action-guiding experiments, namely that they have been performed on a massive scale, mostly in the form of agricultural and technological experiments, thousands of years before the epistemic experiments of modern science. My second thesis is that directly action-guiding experiments have a strong and immediate justification that is not available for epistemic experiments.

5.1 A Very Direct Justification

My problem in vindicating this claim is that it is in a sense self-vindicating. In order to find out whether you can achieve Y by doing X, what better way can there be than to do X and see if Y occurs? In particular, we have no problem in justifying the use of an intervention or manipulation (namely X). We want to know the effects of such an intervention, and then it is much better to actually perform it than for instance to passively observe the workings of nature without performing the intervention. This simple argument, of course, is not available for epistemic experiments.

In the same vein we can easily justify the requirement that directly action-guiding experiments should be repeatable. Since the purpose of the experiment is action-guiding, we need to establish a connection between an intervention that we can perform again and an outcome that will follow after it. Such a connection should appear regularly; it is not sufficient that something happened once.

Someone might counter with the question: Can this be true? Can directly action-guiding experiments really be that strongly justified? Have we not learned that all experiments are theory-laden?

Well, indeed we have, but the discussion on the theory-ladenness of experiments has referred to epistemic experiments, not directly action-guiding ones. The specific theory-ladenness of experiments seems to come with their epistemic interpretations. The following two examples should clarify the difference:

1. An engineer needs an acid-resistant material. She has been recommended two materials. In order to find out which is best she exposes two similar objects, one of each material, to the same strong acid for two months. After that she weighs and inspects the two objects and finds one of them to be much less affected than the other. She concludes that it is the most acid-resistant of the two.
2. A scientist wants to test a hypothesis about chemical bonding. She discovers that according to a well-established theory, if this hypothesis is correct then the addition of a small amount of niobium to iron should make the metal much more resistant to acids. She therefore makes two similar objects, one of each material, and exposes them to the same strong acid for two months. After that she weighs and inspects the two objects and finds the one with niobium to be much less affected than the other. She concludes that the experiment confirmed the hypothesis.

It should be obvious that the second of these experiments, the epistemic one, is theory-laden in several ways that the first is not. Directly action-guiding experiments such as this are in fact remarkably *theory-independent*. You may have whatever reason you want for believing that one piece of metal is more resistant to 2 months in acid than another. When the experiment is finished and you have found one of them to have lost more weight than the other there is not much scope for interpretation of the experiment. The experiment just tells you how it is. This is of course because of the nature of the question that the experiment was constructed to answer. Questions about the effects of practical actions can be answered by performing these actions and monitoring their effects. Questions about the workings, mechanisms, causes, and explanations of natural or social phenomena are different, and the use of experiments to answer them has to be justified by argumentation that of necessity make these experiments[6] much more theory-laden.

Any claim of theory-independence is (among philosophers) a bold statement, so I had better announce some caveats, in fact four of them. First, I am not claiming that directly action-guiding experiments are completely theory-independent. My claim is that they are radically less so than epistemic experiments, and in fact not more theory-dependent than any non-empty statement about empirical subject-matter. Suppose that a craftsman has tried two different mixtures of copper and tin several times and found that one of them yields a much harder bronze than the other. Based on this he concludes that in the future he will get harder bronze with that mixture. This is of course a theory-dependent conclusion since it depends on the theoretical assumption that there are certain types of regularities in nature. However, this is a type of theory-dependence that is difficult to avoid in any statement about what obtains more than once. It is a minimal form of theory-dependence that we cannot

---

[6]Or to be precise: the interpretations of these experiments.

avoid, in contradistinction to the dependence on more specific theories that has been shown to hold for many epistemic experiments.

The second caveat is that the comparative theory-independence of the directly action-guiding experiments does not exclude mistakes in the execution or interpretation of these experiments. You can for instance perform an agricultural experiment badly so that the cultivars to be compared grow under different conditions although you intended their conditions to be the same. You can interpret the same experiment incorrectly by believing its results to be valid under quite different conditions. I will return below to these possible errors and how they can be dealt with.

The third caveat, following rather immediately from the second, is that information from a directly action-guiding experiment does not necessarily override information from other types of investigations. The quality of the investigations will have to be taken into account. As one example of this, suppose that we have strong mechanistic reasons to believe that a proposed new drug has serious side effects on the eyes. Such information can be overridden by information from good clinical trials, but it cannot be overridden by a clinical trial that does not include state-of-the-art ophthalmological examinations of all subjects.

The fourth caveat is that there are cases when directly action-guiding experiments should not be performed in spite of being the best way to obtain the desired knowledge. They may for instance be too expensive or they may be ethically indefensible. Around 1630, Ambroise Paré (c.1510–1590) reportedly performed an experiment in order to find out whether so-called bezoar stones (boli found in the stomachs of goats and cows) are universal antidotes to poisons. The experimental subject was a man convicted to hanging for the theft of two silver dishes. He was promised release if the antidote worked. The poison was administered followed by the antidote, but "he died with great torment and exclamation, the seventh hour from the time that he took the poison being scarcely passed" (Goldstein and Gallo 2001). This was certainly a directly action-guiding experiment, but not one to be commended.

5.2 *The Unwelcomeness of Theory-Independent Information*

Not surprisingly, the presentation of strongly supported, theory-independent information is not always welcomed by the proponents of theories that may have to be revised or deserted due to that information. In subject-areas dominated by strongly held theoretical or ideological views, a person who reports the outcome of a directly action-guiding experiment may find him- or herself in the same situation as the child in Hans Christian Andersen's tale of the Emperor's New Clothes. As you may remember the emperor of the tale continued his pompous parade after being told what was open to everyone to be seen. Academic dignitaries and other defenders of received theories have been known to behave similarly in response to incontrovertible, theory-independent information.

I know of no other field where the unwelcomeness of theory-independent information has been as pronounced as in clinical medicine. It is interesting to compare medicine to agriculture. As we have seen, field trials have been used by farmers since long before modern science. I have not been able to find any record of farmers opposing the method. In contrast, medical treatment experiments had great difficulties in gaining acceptance. Before the twentieth century only few such experiments were performed, and their outcomes were often resisted by the medical authorities. Why this difference between agriculture and medicine? Doctors are not worse educated than farmers, and neither is the outcome of their actions less important. I have found no other plausible explanation of this difference than strongly endorsed theories that prevented doctors from being open to correction by empirical evidence.

An interesting example of this is an experiment on the treatment of pneumonia that was performed in Vienna by the Polish-Austrian physician Joseph Dietl (1804–1878). In the middle of the nineteenth century, the general view among physicians was that pneumonia depends on an imbalance between the bodily fluids. The most commonly recommended treatment was blood-letting. Some physicians favoured instead a somewhat less drastic means of bringing the bodily fluids into balance, namely the administration of an emetic. In 1849 Dietl reported an investigation in which he had compared three groups of pneumonia patients. One group had received blood-letting, the second had received an emetic, and the third had received general care but no specific treatment. Mortality among those who had received blood letting was 20.4 %, among those who had received an emetic 20.7 % and among those who had received no specific treatment only 7.4 %. Dietl's message was at first rather negatively received. His critics claimed that since disease and treatment are highly individual issues, they cannot be settled with statistics. But in a longer perspective he was successful. In the 1870s the major medical textbooks advised against blood-letting of pneumonia patients (Dietl 1849; Kucharz 1981). However, it was only well into the twentieth century that therapeutic experiments became the established method to determine the best treatment for diseases. Physicians were in this respect thousands of years behind farmers.

In the 1930s and 1940s, clinical studies employing control groups became increasingly common in the medical literature. A big step forward was taken in 1948 with the first publication of a clinical trial using modern methodology, including the randomization of patients between treatment groups (Cooper 2011; Doll 1998; Marshall et al. 1948). But the introduction of clinical trials was not without resistance. The opposition has been strongest in psychiatry, which is unsurprising due to the influence in that specialty of entrenched but empirically unsubstantiated doctrines prescribing what treatments to use. But in recent years, it has become generally accepted in psychiatry that the choice between for instance Freudian psychoanalysis and pharmacological treatment should not be based on ideology but instead be made separately for each diagnosis, based on directly action-guiding experiments showing what effects the different treatments have on patients with that diagnosis. Obviously this openness to reality checks is essential for the future progress of psychiatry.

The remaining pockets of resistance in the healthcare sector can now be found among so-called alternative therapists who commonly reject directly action-guiding experiments with arguments very similar to those brought forward by Joseph Dietl's adversaries.

6 The Justification of Epistemic Experiments

As I have already said, the immediate and comparatively theory-independent justification of directly action-guiding experiments is not available for experiments of the epistemic variant. Although this is not my main topic, I should say something about the justification of experiments as a means to gain knowledge of the world we live in. As J.E. Tiles puts it: "How can we possibly learn the principles which govern the action of natural bodies if we do not let nature take its course?" (Tiles 1993).[7]

To begin with we should not expect the justification of epistemic experiments to be as strong as that of directly action-guiding experiments. There are indeed important cases when well-conducted observational but non-experimental studies are the epistemic ideal of practicing scientists, and this for good reason. In zoology, if we want to investigate the actual behaviour of an animal species, our primary sources of knowledge are observations in the wild performed in ways that disturb the animals as little as possible. This is in one respect the extreme opposite of the experimental ideal. In an experiment we intervene strongly enough to determine as exactly as possible what will happen. In a zoological field study we intervene as little as possible. The same applies to many other types of studies in both the natural and the social sciences. In brief, experimentation is unsuitable when we want to know how something develops spontaneously. We use it, however, in a wide range of circumstances when we are looking for regularities and mechanisms. A discussion of the epistemic usefulness of experiments should be limited accordingly. What we need to explain is our use of experiments only for a class of knowledge claims that can at least roughly be specified as knowledge that pertains to regularities and mechanisms.

A necessary assumption of experimentation is that the same regularities ("laws") that govern the spontaneous workings of nature also apply when nature responds to human intervention. This was assumed by Francis Bacon (1561–1626), who also went one step further and claimed that we can only learn the regularities of nature if we expose it to the extreme conditions of experimentation: "For like as a man's disposition is never well known till he be crossed, nor Proteus ever changed shapes till he was straitened and held fast; so the passages and variations of nature cannot appear so fully in the liberty of nature as in the trials and vexations of art." (Bacon, [1605] 1869, 90; cf. Zagorin 1998, 61–62)

One interesting explanation of the usefulness of experiments was provided by John Stuart Mill in his *System of Logic*:

> For the purpose of varying the circumstances, we may have recourse (according to a distinction commonly made) either to observation or to experiment; we may either find an instance in nature suited to our purposes, or, by an artificial arrangement of circumstances, make one. The value of the instance depends on what it is in itself, not on the mode in

---

[7]An interesting criticism of experimental methodology can be found in Thomas Hobbes (Shapin 1985, 1996, 110–111).

which it is obtained... Thus far the advantage of experimentation over simple observation is universally recognised: all are aware that it enables us to obtain innumerable combinations of circumstances which are not to be found in nature, and so add to nature's experiments a multitude of experiments of our own. (Mill [1843] 1974, 381–382)

This is a rather diffident justification of experimentation. Mill does not defend the specific characteristic of experiments, namely the active element, the intervention that distinguishes it from mere observations. He only defends it in terms of an advantage that experiments often but not always have over observations, namely the advantage of providing data from a much larger number of combinations of circumstances. There are research areas where experiments usually do not have this advantage since they are expensive and difficult to perform. With this justification it would be difficult to justify experiments in those areas.[8] We must therefore ask: Is there any justification of the epistemic use of experiments that specifically justifies its characteristic component, namely intervention?

In my view there is. In order to explain it I first have to introduce the view on causation that it is based on.

6.1 Two Notions of Causality

We can distinguish between two meanings of causality. First, it can refer to cause-effect relationships. These are binary production relationships such that if C is a cause of the effect E, then, in the absence of contravening circumstances, if C takes place then so does E. Secondly, by causality we can mean the totality of regularities in the universe, or its workings. These two senses are often taken to coincide. To know how something works and to know the cause-effect relationships that determine its operations would seem to be one and the same thing.

However, it is not the same thing. The idea that the workings of the universe consist in binary cause-effect relationships is a model of reality, and moreover it is a rather rough model that works well in some circumstances but not in others. This was seen clearly by Bertrand Russell, who observed that "oddly enough, in advanced sciences such as gravitational astronomy, the word 'cause' never occurs" (Russell 1913, 1). Furthermore, he said:

In the motions of mutually gravitating bodies, there is nothing that can be called a cause and nothing that can be called an effect: there is merely a formula. Certain differential equations can be found, which hold at every instant for every particle of the system, and

---

[8]Mill adds a second justification, namely that "[w]hen we can produce a phenomenon artificially, we can take it, as it were, home with us, and observe it in the midst of circumstances with which in all other respects we are accurately acquainted." We can for instance produce in the laboratory, "in the midst of known circumstances, the phenomena which nature exhibits on a grander scale in the form of lightning and thunder". (382) This is also a very modest defence of experimentation, since it only applies to the phenomena that do not occur spontaneously under the type of circumstances that we are well acquainted with but can nevertheless be transferred to such circumstances.

which, given the configuration and velocities at one instant, or the configurations at two instants, render the configuration at any other earlier or later instant theoretically calculable. (Russell 1913, 14)

The cause-effect pattern would have been sufficient for a full description of a clockwork universe in which all motions are produced directly by some other movement. This is the type of mechanistic model that can be found for instance in the natural philosophies of Robert Boyle (1627–1691) and René Descartes (1596–1650), who assumed that natural phenomena can be described in terms corresponding to how "the movements of a clock or other automaton follow from the arrangement of its counter-weights and wheels."[9] However, this is a pre-Newtonian model of the universe. The cause-effect pattern does not capture Newtonian physics, in which movements emerge from complex interactions among a large number of bodies, all of which influence each other simultaneously.

Modern physics is of course even further away from the cause-effect model than the physics that Russell referred to. Furthermore, several other sciences have followed physics in adopting models in which the flow of events is determined by simultaneous mutual influences that are describable by systems of equations, rather than by the stepwise production of effects in a causal chain. This applies for instance to climatology, economics, and biological population dynamics. In all these areas, an account restricted to binary cause-effect relationships will lack much of the explanatory power of modern science.

Of course, the cause-effect model is useful for many purposes. However, this does not mean that the development of the world 'consists' of cause-effect relationships, only that the cause-effect model is a useful approximation in these cases.

Why then do we so consistently refer to cause-effect relationships? The answer to that question is much easier to give once we have made the distinction I just mentioned between two notions of causality. Our answer should not be an attempt to find out how things really are, what the workings of the universe are. Instead, what we need to explain is why we so strongly prefer the cause-effect model in spite of its insufficiency.

The answer I propose is that we base our understanding of the world on thought patterns that accord with our experiences of our own interactions with it. I hit the nail with the hammer, and the nail is driven into the plank, hence my act causes the movement of the nail. Since this occurs regularly, I can use hammer-blows as means to achieve ends consisting in having nails infixed in planks. Based on this understanding of our own interactions with the world we tend to explain what happens without our agency in analogous ways. We throw a stone into a pond, producing ripples on the surface. Then we see a twig falling from a tree into the pond, followed by similar ripples. We conclude that the same type of relationship, a cause-effect relationship, is in operation in both these sequences of events.

---

[9] ...les mouvements d'une horloge, ou autre automate, de celle de ses contrepoids et de ses roués (Descartes [1632] 1987, 873).

This is a version of the interventionist theory (manipulability theory, agency theory) of causality. Its basic assumption is that we acquire the notion of causation from our experiences as agents (Price 1992, 514). But note that contrary to some other interventionist accounts, this one does not refer to the workings of the universe. Therefore it is not sensitive to the most common argument against interventionist notions of causality, namely that they represent an unduly anthropocentric view of the universe. What I have presented is an anthropocentric view, but not of the universe but of a thought model that seems to be deeply entrenched in our ways to think about the universe.

6.2 *Experiment and Causation*

In summary: although the cause-effect model of the universe is problematic, we have a strongly entrenched tendency to employ it. When trying to understand the world we want to do so in terms of cause-effect relationships. This is where experiments come in. Experiments are constructed to show how a certain intervention produces a certain outcome. The relationship between intervention and outcome is fairly close to a cause-effect relationship. This cognitive fit, I propose, is a major reason why experiments tend to be so useful in our strivings to understand the world.

There is an obvious counterargument, or source of disappointment, that can be expected in response to this justification of experimentation: It does not justify experimentation as the best way for a perfectly rational being to investigate the world. Instead it justifies experimentation for us humans as we are, warts and all, with the cognitive limitations that prevent us from being completely rational. Should not philosophy go beyond such limitations and concern itself with perfect rationality?

My answer is no. Whether we want it to be so or not, philosophy is a human enterprise, concerned with the conditions of the human race rather than with our (of necessity rather naïve) musings about how some entity without these limitations would or could think. I do not believe that we can rid ourselves of our proclivity to think of the world in terms of causes and effects. Seen in that perspective, the conformity of the experimental method to cause-effect thinking is a legitimate justification of the epistemic use of experiments, albeit a justification with restrictions and limitations.

7 Problems, Limitations, and Remedies

I will now move on to the third thesis about directly action-guiding experiments. It refers to the sources of error in experiment. As I have already emphasized, the strong and immediate justification of directly action-guiding experiments does not protect us against mistakes in the execution and interpretation of these experiments.

Like almost everything else they can be performed badly, and in particular they can be subject to wishful thinking and other cognitive failures. The third thesis is that the safeguards needed to avoid mistakes are essentially the same in action-guiding and epistemic experiments, and that these safeguards have to a large extent been developed for action-guiding experiments. To a significant degree, science learned from technological experimentation how to perform experiments properly.

I will show this by considering in turn six important safeguards in experimental design against errors of execution and interpretation:

- Control experiments
- Parameter variation
- Outcome measurement
- Blinding
- Randomization
- Statistical evaluation

7.1 Control Experiments

According to the simple recipe for an action-guiding experiment that I have been referring to, we perform X in order to find out whether Y then occurs. In some cases this can be done against the background knowledge that without X, Y will not occur. But if such knowledge is not available, or can be put to doubt, then we need to perform a control experiment: We need to refrain from performing X, or perform some relevant alternative to X, in order to see whether Y will then happen. In other words we need to perform a control experiment or, as it is often called, a control arm of the experiment.

Control experiments are needed in epistemic experiments for quite similar reasons. Suppose that we find Y occurring after introducing X, and conclude that X has some role in the mechanism giving rise to Y. If it turns out that Y also occurs without X, or with some alternative to X, then we seem have drawn a too rash conclusion. A concrete example: Suppose that I find out that a certain chemical solution changes its colour from blue to red if I stir it with a copper object, and use this information in support of some chemical theory about reactions with copper. I would have to revise my view if it turns out that the same colour change takes place with a wooden spoon or without any stirring at all.

Some of the examples mentioned above show us that the use of controls goes back to pre-scientific, directly action-guiding experiments. The Andean farmers tried out different crops in their fields, and these were clearly comparative experiments. The same applies to the hungoo of the Mende people in Sierra Leone in which different agricultural practices were compared in adjacent parts of a field. An early description of a directly action-guiding experiment with a control arm can be found in the Book of Daniel, one of the books of the Hebrew Bible (written around 167 BCE). In its first chapter, the Babylonean king Nebuchadnezzar II (who reigned

c. 605–562 BCE) is said to have endorsed an experiment in which a group of men were allowed to live for 10 days on vegetables and water instead of the meat and wine that were the regular diet of the King's men. At the end of the trial, "their countenances appeared fairer and fatter in flesh than all the children which did eat the portion of the king's meat."[10] Consequently they were allowed to continue on their vegetarian diet.

The most famous early treatment experiment that compared different treatments in a systematic way is James Lind's scurvy trial in 1747. As surgeon of the British naval ship Salisbury he had an increasing number of patients with scurvy. He divided twelve afflicted sailors into six groups of two. Each of the six groups received a different treatment: cider, a weak acid, vinegar, seawater, nutmeg and barley water, or oranges and lemons. After 6 days the two men on a citrus treatment had regained health, whereas all the others remained sick (Collier 2009).

Today control groups are a self-evident part of the design of medical treatment experiments. The same applies to directly action-guiding experiments trying out methods in social work or education, but in those areas their introduction came later than in medicine. The first control group experiment in education seems to have been a study reported in 1907 by J.E. Coover and Frank Angell (Dehue 2005). They were interested in whether training of certain skills improves the performance of other abilities. Such effects had been obtained in a previous experiment, but they found it "to be regretted" that the authors of the previous study "did not carry on a 'control' experiment along with their tests to ascertain the training effect of the tests themselves and to throw additional light on the changes taking place in the training intervals" (Coover and Angell 1907, 329). In their own experiment they compared subjects receiving training to subjects not doing so, and obtained evidence for instance that training in the discrimination of intensities of sounds improved the ability to discriminate between shades of grey.

Examples of control arms in epistemic experiments are also easily found.[11] They are a standard component of experimental procedures across all disciplinary borders. What I hope to have shown is that the use of controls predates science and is indeed part of what it took over from technological experimentation.

---

[10]Book of Daniel 1:15, King James Version.

[11]The earliest example I am aware of was reported from China in the third century CE. A woman was accused of having murdered her husband and thereafter burning down the house with his body. She claimed that he had burned to death in a fire. The magistrate ordered one dead and one living pig to be burned in a shed. The pig burned alive had a lot of ashes in its mouth whereas the pig previously killed had none. Since the dead man had no ashes in his mouth this was taken as evidence that he had not been burned alive (Lu and Needham 1988).

7.2 Parameter Variation

One of the standard features of modern scientific experiments is parameter variation. By this is meant that the experiment is performed repeatedly with systematic variation of one or several variables. For instance, if we wish to determine the effects of temperature on a chemical reaction, we perform the experiment at different temperatures. If we wish to determine the effects of a supposed catalyst, we perform the experiment both with and without the substance. Parameter variation can be seen as a generalization of the control group. Instead of looking for Y both with and without some X ("without" being the control experiment) we look for it with many variants of X.

The term "parameter variation" usually refers to numerical variables, but the principle is also applicable to non-numerical variation. This can be illustrated with Galileo's experiment on floating bodies. In order to show that it does not depend on a body's form whether it will sink or float in water, he performed an experiment on objects of different shape but with the same specific weight. In explaining the experiment, Galileo emphasized the usefulness of changing only one variable at a time.

> For were we to make use of materials that could vary in specific weight from one to another, when we encountered variation in the fact of descent or ascent we would always remain with ambiguous reasoning as to whether the difference derived truly from shape alone, or also from different heaviness. (Drake 1981, 74)

Alhazen (965–c.1040) performed a famous optical experiment in which he investigated the effects of varying the size of an aperture through which moonlight was projected. Based on this experiment he has been acclaimed as the first to make "systematic use of the principle of persistent variation of the experimental conditions" (Schramm 1963, 287), in other words to have made the first experiment with parameter variation. Without in any way depreciating Alhazen's great and often sadly neglected contributions to science, it should be recognized in fairness that such parameter variation was a common feature of pre-scientific, directly action-guiding experiments. As mentioned above, experiments with systematic variation of the proportions of major ingredients in glass were performed in Raqqa in the early Islamic period, and we have reasons to believe that earlier than that, metallurgic batch recipes were developed in the same way. In other words, parameter variation was developed in pre-scientific, technological experiments. It can be hypothesized that just like the control experiment, it was part of what science took over from technological experimentation.

7.3 Outcome Measurement

Measurement of the outcome is usually not mentioned in accounts of the safeguards necessary to protect us against errors in experiments and experiment-based reasoning. However, in both directly action-guiding and epistemic experiments it is essential to measure the outcome as exactly as possible whenever such measurement is possible. Of course there may be aspects of the outcome that we cannot measure. A farmer trying a new cultivar will be interested in features of the plant such as taste that cannot measured. But he will also be interested in the yield that can be measured for instance with a suitable container. In order to compare the yields from two parts of the same field it may be sufficient to just collect the harvests separately and look at them. But in order to compare yields from different fields, or different years, such visual impressions will be too uncertain, and numerical measurement appears necessary.

In epistemic experiments, measurement fills the same function as in directly action-guiding experiments. In addition, measurement provides the means necessary to determine numerical relationships that can be compared to mathematical models of the workings of nature. Historically, physics pioneered the use of outcome measurement in scientific experimentation, as can be seen in the work of forerunners such as Alhazen and Galilei. Measurement, in particular of weight and length, was well established for technical and other purposes long before modern science. It is also plausible that such devices were used in technological experiments. However, due to the lack of documentation we do not know if that was really the case. What we do know, however, is that the methods of measurement used in scientific experimentation were largely based on measurement techniques that had been developed for technological purposes.

7.4 *Blinding*

Our expectations on an experiment can influence its outcome in at least three major ways. Think again of a farmer who tries out two different cultivars. He grows them on arid land and intends to give them the same amounts of water. Suppose that he initially believes one of the two cultivars to resist draught much better than the other. These expectations can influence the experiment in two ways.

First, when watering the two fields he may fail to give them equal amounts of water. In other words his behaviour can be influenced by his expectations. Secondly he can be misled by his own preconceptions when inspecting the plants, and perhaps believe the cultivar he favours to be somewhat more vigorous although an impartial observer would see no difference. In other words his expectations can influence his evaluation. The same two effects can also occur when craftspeople perform an experiment with two different tools in order to see which yields the best result. If the person who actually operates the tool has expectations on the outcome, then that can lead her to (unconsciously) work harder or better with one of the tools than with the other. If the person evaluating the outcome (who may or may not be the same person) has expectations about the outcome, then these expectations can influence her judgment.

5 Experiments Before Science

These two mechanisms are virtually ubiquitous in both directly action-guiding and epistemic experiments. This is because in the vast majority of cases, there is someone whose actions are part of the performance of the experiment, and there is also someone who judges the outcome.

In some experiments there is also a third type of mechanism. Suppose that I am one of the experimental subjects in a trial of two painkillers. If I believe one of them to be more efficient than the other, then chances are high that I will report less pain when taking it than when taking the other, since expectations have an influence on pain. This is an expectation effect on the experimental subject. Contrary to the other two expectation effects it only appears in studies with humans as experimental subjects.

Various methods can be used to reduce the three expectation effects. Measurement of the outcome, for instance, helps against the expectation effects on outcome evaluation. But there is one method that surpasses all others in neutralizing all three types of expectation effects, namely *blinding* (also called masking). By this is meant that the persons concerned are left ignorant about that which they might otherwise have had expectations about. In our examples, the person who waters the two crops should not know which is which, and neither should the person who judges the quality of the plants. Similarly, the person who operates the two tools should if possible not know which is which, and neither should the person who judges the outcomes of her work. In the experiment with the two painkillers, there are three possible expectation effects and therefore three categories of persons who should be kept ignorant of which drug each of the experimental subjects receives: the experimental subjects themselves, the physician who prescribes the pills and recommends the subjects to take them, and the physician who interviews the subjects about the outcome. (Since it is common for one and the same physician to have the last two roles, such threefold blinding is commonly called "double blinding".)

The history of observational and experimental blinding remains to be written. It has been claimed that Dom Pérignon (1639–1715), a French monk and wine maker, performed blind-testing of wines in order not to be influenced by expectation effects (Bullock et al. 1998), but I have not been able to verify that claim in reliable historical sources. In 1817, Stradivarius violins were compared to other violins in blind hearing tests (Fétis 1868; Quatremère de Quincy 1817, 249). The earliest scientific study with blinded evaluators seems to have been that performed in 1784 by a commission of the French Academy of Sciences, led by Benjamin Franklin, that investigated Franz Mesmer's claims of animal magnetism. Under blinded conditions, mesmerists were unable to distinguish which objects had gone through an occult procedure described as filling them with vital fluid. The subjects reported presence of such fluid when they had been led by deception to believe that there was "fluid" in some place although the procedure supposed to produce it had not taken place. Conversely, when they were led to believe that there was no fluid, but the mesmerist was indeed performing the procedure supposed to produce it, they did not report any fluid. The commission concluded from these blinded experiments that the fluid had no physical existence. The alleged effect could be completely accounted for as an expectation effect on the evaluators (Lopez 1993; Sutton 1981). Blinding was also used in studies performed in 1799 by the British chemist Humphry Davy. When testing the effects of nitrous oxide (laughing gas) he kept the subjects unaware of whether they were exposed to the gas or not (Holmes 2008).

Blinding was taken up by critics of deviant belief systems and used rather extensively in the nineteenth century to disclose scams and self-deception. One example is the experiments performed in the 1830s by Michel Eugène Chevreul (1786–1889) on the use of a pendulum as a dowsing device. Initially he found that a pendulum held by himself gave a clear indication when he placed it directly over mercury but not when a glass pane was inserted over the mercury. He then repeated the experiment while blind-folded, letting an assistant introduce the glass pane in some of the trials, without telling Chevreul when he did so. The dowsing effect completely disappeared and so, unsurprisingly, have all dowsing effects done since then when investigated under experimental conditions including efficient blinding (Zusne and Jones 1982, 249–255). Similarly, in 1835 a double blinded (and randomized) test was performed in Nuremberg, in which a homeopathic drug was compared to pure water. No effect of the homeopathic drug was found (Stolberg 2006).

For a long time, blinding was with few exceptions only used to expose effects that were supposed to be entirely due to suggestion or fraud. At the very end of the nineteenth century several researchers began to use it as a means to improve the accuracy of observations in experiments where "real" effects were expected. After World War II awareness of the fallibility of human judgments became widespread among researchers, not least due to influence from psychological research. As a consequence of that, blinding became generally accepted as a means to achieve reliable observations (Kaptchuk 1998). The use of blinding has been standard in clinical trials since the 1940s, and "randomized double-blinded" is now the gold standard in that field. In many other areas where blinding can reduce the risk of experimenter error it is still not the standard procedure.

Summarizing all this, blinding – or at least its systematic use – differs from some of the other safeguards in not being part of what scientific experimenters learned from technological experiments. This technique has been developed within science. Arguably, its use is best developed in clinical trials that are of course directly action-guiding experiments performed by scientists.

7.5 Randomization

One of biases that may effect experiments is selection bias. Suppose that you test different cultivar on various test fields, and find one of them to yield better harvests. An alternative explanation can be that the field selected for that cultivar had better soil than the others. Similarly, if a drug is tested on two groups of patients, the results are not of much value if the most successful drug was given to the group that had on average better health than the other. Various measures can be taken to avoid this source of experimental error. The method usually preferred is randomization, i.e. letting chance decide instead of a researcher who can always, consciously or unconsciously, be influenced by her own expectations.

Randomization appears to be a rather modern phenomenon. Most of the development has taken place in medicine. Interestingly, three modes of randomization have been proposed. The first of them is to first pick out the groups to be treated in the experiment, and then randomize the treatments between these groups. This method was proposed by Jan Baptist van Helmont (1580–1644) in a challenge concerning the efficacy of different medical treatments:

> Let us take out of the hospitals, out of the Camps, or from elsewhere, 200, or 500 poor People, that have Fevers, Pleurisies, &c. Let us divide them into halfes, let us cast lots, that one half of them may fall to my share, and the other to yours; ... we shall see how many funerals both of us shall have: *But* let the reward of the contention or wager, be 300 florens, deposited on both sides. (quoted from Armitage 1982)

This method has an obvious disadvantage: Although it equalizes the chances of the two combatants, it does not necessarily lead to a usefully action-guiding outcome. If the two groups of patients differ so that one is on average in worse health than the other, then the initial lottery may have a larger influence on the outcome than the difference between the two treatments. Helmont's challenge was not accepted, and as far as I know this method is just a historical curiosity that achieved no following.

A much better strategy is alternation (alternate allocation), by which is meant that the first, third, fifth etc. patients included in the trial are assigned to one group and the second, fourth, sixth etc. to another. Alternation was described as early as 1816 in a PhD thesis on blood-letting by Alexander Hamilton at the University of Edinburgh. It is unclear whether the trial described in that thesis actually took place, but the method seems to have been used on at least a few occasions in nineteenth century medical experiments. In the 1920s it was quite common (Chalmers 2001).[12] However, as pointed out by Richard Doll, it had problems:

> The technique of alternate allocation had one major disadvantage: the investigator knew which treatment the next patient was going to receive and could be – and indeed often was – biased by knowing what the next treatment would be when deciding whether or not a patient was suitable for inclusion in the trial. (Doll 1998, 1217)

The method that was eventually chosen, and is now standardly used, consists in randomizing each participant to one of the groups. This method avoids the disadvantages of the other two methods. Interestingly though, it did not originate in medicine. Instead it was developed in the context of experimental agriculture

---

[12] It is commonly claimed that the Danish physician Johannes Fibiger (1867–1928) used it in a trial of diphtheria treatment in 1898. However, in that study yet another, more uncommon method was used: Patients admitted on days 1, 3, 5 etc. of the trial were assigned to one group and those admitted on days 2, 4, 6 etc. to the other (Fibiger 1898; Hróbjartsson et al. 1998).

by the statistician Ronald Fisher (1890–1962) in the 1920s when working at the Rothamsted Experimental Station in England. He developed statistical methods for agricultural trials that included the random assignment of cultivars to fields. The method began to be used in clinical medicine in the late 1940s. In 1948 the first medical study employing the method was published (Doll 1998; Marshall et al. 1948). Since then it has spread to a large number of other research areas, including psychology, the social sciences, and experimental biology (where animals rather than humans are distributed randomly between treatment groups). As should be clear from this brief history, randomization appears to have been developed entirely within scientific rather than pre-scientific experimentation. Just as for blinding, scientific directly action-guiding experiments had a major role in these developments.

7.6 Statistical Evaluation

With randomization and other safeguards we can diminish the effects of pure chance on experimental results. However, we cannot eliminate them completely. Suppose that a clinical trial shows a small difference between patients receiving two treatments. We must then determine whether that effect is so small that it can reasonably have been a chance effect. The same problem appears in virtually all disciplines in which experiments are performed, irrespective of whether they are epistemic or directly action-guiding. Statistical methods developed in the last 100 years or so have made it possible to deal with this problem in a systematic way. These methods can – and should – be adjusted to the evidence requirements ("burden of proof") that are appropriate in the particular issue at hand. For this final item on our list of safeguards, the historical evidence is quite clear and in no need of being treated at length: These tools were not available in prescientific experimental traditions. They have been developed within science.

7.7 Summary

The available evidence gives a fairly clear picture of the six safeguards. There can hardly be any doubt that they are all needed in both directly action-guiding and epistemic experiments. Three of them, namely control experiments, parameter variation, and outcome measurement appear to have been parts of the methodologies and ways of thinking that scientific experimentation took over from technological experiments. The other three, namely blinding, randomization, and statistical evaluation, have their origin in modern science. In conclusion, technological experiments that preceded science have contributed substantially to the experimental methodology of modern science.

8 Conclusion

I hope to have shown that directly action-guiding experiments are an important category that should be kept apart from epistemic experiments. Furthermore, I hope to have substantiated my three theses about directly action-guiding experiments, namely:

- The origin of experimentation is neither academic nor curiosity-driven. Instead, it is socially widespread and driven by practical needs. In particular, technological experiments predated scientific experiments, probably with thousands of years.
- Action-guiding experiments have a strong and immediate justification and are as theory-independent as empirical statements can be. In this they differ from epistemic experiments.
- The safeguards needed to avoid mistakes in the execution and interpretation of experiments are essentially the same for action-guiding and epistemic experiments. Several of these safeguards are parts of the heritage from technological experiments that science has taken over.

Directly action-guiding experiments are a large part, arguably the vast majority, of the experiments we humans perform and have performed. I have discussed the ancient tradition of such experiments, but I have said nothing about currently ongoing experiments of the same nature in workshops, farms and many other places. They are an interesting area of study, for social scientists and philosophers alike. The American anthropologist Allen Johnson has asked a question that I believe to be of considerable interest for philosophers as well: "To what extent is experimentation a characteristic of all domains of human behavior? In what spheres is such experimentation conducted openly, as in the case of agriculture, and in what spheres is it conducted surreptitiously, behind a mask of conformist ideology?" (Johnson 1972, 157–158).

Directly action-guiding experiments, both within and outside of science, are philosophically interesting in their own right, not least due to their role in practical rationality. In addition they are a necessary background for our understanding of the other major type of experiments, namely those that are undertaken for epistemic rather than practical goals. I doubt that we can properly understand the epistemic role of experiments in science without relating and comparing them to the directly action-guiding, mainly technological, experiments that they originated from. This, by the way, is one of the many reasons why the philosophy of science cannot do without the philosophy of technology.

References

Alcorn, J. B., & Toledo, V. M. (1998). Resilient resource management in Mexico's forest ecosystems: The contribution of property rights. In F. Berkes, & C. Folke (Eds.), *Linking social and ecological systems. Management practices and social mechanisms for building resilience* (pp. 216–249). Cambridge: Cambridge University Press.

Armitage, P. (1982). The role of randomisation in clinical trials. *Statistics in Medicine, 1*, 345–352.

Bacon, F. ([1605] 1869). In W. A. Wright (Ed.), *The advancement of learning [Of the proficience and advancement of learning, divine and human]*. Oxford: Clarendon Press.

Berkes, F., Johan C., & Folke, C. (2000). Rediscovery of traditional ecological knowledge as adaptive management. *Ecological Applications, 10*, 1251–1262.

Biggs, S. D. (1980). Informal R&D. *Ceres, 13*(4), 23–26.

Bray, W. (2000). Ancient food for thought. *Nature, 408*(9), 145–146.

Bullock, J. D., Wang, J. P., & Bullock, G. H. (1998). Was Dom Perignon really blind? *Survey of Ophthalmology, 42*, 481–486.

Chalmers, I. (2001). Comparing like with like: Some historical milestones in the evolution of methods to create unbiased comparison groups in therapeutic experiments. *International Journal of Epidemiology, 30*(5), 1156–1164.

Chandler, P. M. (1991). The indigenous knowledge of ecological processes among peasants in the People's Republic of China. *Agriculture and Human Values, 8*, 59–66.

Claridge, J. A., & Fabian, T. C. (2005). History and development of evidence-based medicine. *World Journal of Surgery, 29*, 547–553.

Collier, R. (2009). Legumes, lemons and streptomycin: A short history of the clinical trial. *Canadian Medical Association Journal, 180*(1), 23–24.

Cooper, M. (2011). Trial by accident: Tort law, industrial risks and the history of medical experiment. *Journal of Cultural Economy, 4*(1), 81–96.

Coover, J. E., & Angell, F. (1907). General practice effect of special exercise. *American Journal of Psychology, 18*, 328–340.

Dehue, T. (2005). History of the Control Group. In B. S. Everitt & D. C. Howell (Eds.), *Encyclopedia of statistics in behavioral science* (pp. 2:829–836). Chichester: Wiley.

DeLaine, J. (1990). Structural experimentation: The lintel arch, corbel and tie in western Roman architecture. *World Archaeology, 21*(3), 407–424.

Descartes, R. ([1632] 1987). *Traité de l'Homme*. In René Descartes, *Oeuvres et lettres. Textes présentés par André Bridoux*. Paris: Gallimard.

Dietl, J. (1849). *Der Aderlass in der Lungenentzündung*. Wien: Kaulfuss Witwe, Prandel & Comp.

Doll, R. (1998). Controlled trials: The 1948 watershed. *BMJ: British Medical Journal, 317*(7167), 1217–1220.

Doyon-Bernard, S. J. (1990). From twining to triple cloth: Experimentation and innovation in ancient Peruvian weaving (ca. 5000–400 BC). *American Antiquity, 55*, 68–87.

Drake, S. (1981). *Cause, experiment, and science: A Galilean dialogue, incorporating a new English translation of Galileo's Bodies that stay atop water, or move in it*. Chicago: University of Chicago Press.

Earls, J. (1998). The character of Inca and Andean agriculture. Essay available at http://macareo.pucp.edu.pe/~jearls/documentosPDF/theCharacter.PDF.

Eastwood, B. S. (1968). Mediaeval empiricism: The case of Grosseteste's optics. *Speculum: A Journal of Mediaeval Studies, 43*, 306–321.

Edgeworth, R., Dalton, B. J., & Parnell, T. (1984). The pitch drop experiment. *European Journal of Physics, 5*, 198–200.

Fétis, F.-J. (1868). *Biographie Universelle des Musiciens et Bibliographie Générale de la Musique*, Tome 1 (2nd ed.). Paris: Firmin Didot Frères, Fils, et Cie.

Fibiger, J. (1898). Om Serumbehandling af Difteri. *Hospitalstidende, 6*(12), 309–325; 6(13), 337–350.

Goldstein, B. D., & Gallo, M. A. (2001). Paré's law: The second law of toxicology. *Toxicological Sciences, 60*, 194–195.

Habermas, J. (1968). *Erkenntnis und Interesse*. Frankfurt am Main: Suhrkampf Verlag.

Habermas, J. (1978). *Knowledge and Human Interests* (J. J. Shapiro (Trans.), 2nd ed.). London: Heinemann.

Hansson, S. O. (2014). Beyond experimental philosophy. *Theoria, 80*, 1–3.

Henderson, J., McLoughlin, S. D., & McPhail, D. S. (2004). Radical changes in Islamic glass technology: Evidence for conservatism and experimentation with new glass recipes from early and middle Islamic Raqqa, Syria. *Archaeometry, 46*(3), 439–468.

Henderson, J., Challis, K., O Hara, S., McLoughlin, S., Gardner, A., & Priestnall, G. (2005). Experiment and innovation: Early Islamic industry at al-Raqqa, Syria. *Antiquity, 79*(303), 130–145.

Herschel, W. (1831). *A preliminary discourse on the study of natural philosophy*, part of Dionysius Lardner (Cabinet cyclopaedia). London: Longman.

Holmes, R. (2008). *The age of wonder: How the romantic generation discovered the beauty and terror of science*. London: Harper.

Hróbjartsson, A., Gøtzsche, P. C., & Gluud, C. (1998). The controlled clinical trial turns 100 years: Fibiger's trial of serum treatment of diphtheria. *BMJ: British Medical Journal, 317*(7167), 1243–1245.

Jevons, W. S. (1920). *The principles of science: A treatise on logic and scientific method*. London: Macmillan.

Johnson, A. W. (1972). Individuality and experimentation in traditional agriculture. *Human Ecology, 1*(2), 149–159.

Kaptchuk, T. J. (1998). Intentional ignorance: A history of blind assessment and placebo controls in medicine. *Bulletin of the History of Medicine, 72*(3), 389–433.

Klein, U. (1996). Experiment, Spiritus und okkulte Qualitäten in der Philosophie Francis Bacons. *Philosophia Naturalis, 33*(2), 289–315.

Klein, U. (2005). Experiments at the intersection of experimental history, technological inquiry, and conceptually driven analysis: A case study from early nineteenth-century France. *Perspectives on Science, 13*, 1–48.

Kucharz, E. (1981). The life and achievements of Joseph Dietl. *Clio Medica. Acta Academiae Internationalis Historiae Medicinae Amsterdam, 16*(1), 25–35.

Lelas, S. (1993). Science as technology. *British Journal for the Philosophy of Science, 44*(3), 423–442.

Lopez, C.-A. (1993). Franklin and Mesmer: an encounter. *Yale Journal of Biology and Medicine, 66*(4), 325–331.

Lu, G.-D., & Needham, J. (1988). A history of forensic medicine in China. *Medical History, 32*(4), 357–400.

Malina, J. (1983). Archaeology and experiment. *Norwegian Archaeological Review, 16*(2), 69–78.

Mark, R. (1972). The structural analysis of Gothic cathedrals. *Scientific American, 227*(5), 90–99.

Mark, R. (1978). Structural experimentation in Gothic architecture: Large-scale experimentation brought Gothic cathedrals to a level of technical elegance unsurpassed until the last century. *American Scientist, 66*(5), 542–550.

Marshall, G. et al. (1948). Streptomycin treatment of pulmonary tuberculosis. A medical research council investigation. *British Medical Journal, 2*(4582), 769–782.

McEvoy, J. (1982). *The philosophy of Robert Grosseteste*. Oxford: Clarendon Press.

McGinnis, J. (2003). Scientific methodologies in Medieval Islam. *Journal of the History of Philosophy, 41*, 307–327.

Mill, J. S. ([1843] 1974). *A system of logic ratiocinative and inductive*. In J. M. Robson (Ed.), *Collected works of John Stuart Mill* (Vol. VII). Toronto: University of Toronto Press.

Moropoulou, A., Bakolas, A., & Anagnostopoulou, S. (2005). Composite materials in ancient structures. *Cement & Concrete Composites, 27*, 295–300.

Pesic, P. (1999). Wrestling with Proteus: Francis Bacon and the 'Torture' of Nature. *Isis, 90*, 81–94.

Price, H. (1992). Agency and causal asymmetry. *Mind, 101*, 501–520.

Prioreschi, P. (1994). Experimentation and scientific method in the classical world: Their rise and decline. *Medical Hypotheses, 42*(34), 135–148.

Quatremère de Quincy, A. C. (1817). Institut de France. *Le Monietur Universel* (22 Aug 1817, Vol. 234, p. 924).

Richards, P. (1986). *Coping with Hunger. Hazard and experiment in an African rice-farming system*. London: Allen & Unwin.

Richards, P. (1989). Farmers also experiment: A neglected intellectual resource in African science. *Discovery and Innovation, 1*, 19–25.

Robison, W. (2008). Hume and the experimental method of reasoning. *Southwest Philosophy Review, 10*(1), 29–37.

Russell, B. (1913). On the notion of a cause. *Proceedings of the Aristotelian Society, 13*, 1–26.

de Schlippe, P. (1956). *Shifting cultivation in Africa. The Zande system of agriculture*. London: Routledge & Kegan Paul.

Schramm, M. (1963). *Ibn al-Haythams Weg zur Physik*. Wiesbaden: Franz Steiner Verlag.

Shapin, S. (1985). *Leviathan and the air-pump: Hobbes, Boyle, and the experimental life*. Princeton: Princeton University Press.

Shapin, S. (1996). *The scientific revolution*. Chicago: University of Chicago Press.

Snively, G., & Corsiglia, J. (2000). Discovering indigenous science: Implications for science education. *Science Education, 85*, 6–34.

Stolberg, M. (2006). Inventing the randomized double-blind trial: The Nuremberg salt test of 1835. *Journal of the Royal Society of Medicine, 99*, 642–643.

Sutton, G. (1981). Electric medicine and mesmerism. *Isis, 72*(3), 375–392.

Tiles, J. E. (1993). Experiment as intervention. *British Journal for the Philosophy of Science, 44*(3), 463–475.

Winston, A. S., & Blais, D. J. (1996). What counts as an experiment?: A trans-disciplinary analysis of textbooks, 1930–1970. *American Journal of Psychology, 109*, 599–616.

Wolfe, M., & Mark, R. (1974). Gothic cathedral buttressing: The experiment at Bourges and its influence. *Journal of the Society of Architectural Historians, 33*(1), 17–26.

Zagorin, P. (1998). *Francis Bacon*. Princeton: Princeton University Press.

Zilsel, E. (1941). The origin of William Gilbert's scientific method. *Journal of the History of Ideas, 2*, 1–32.

Zilsel, E. (1942). The sociological roots of science. *American Journal of Sociology, 47*, 544–562.

Zilsel, E. (2000). *The social origins of modern science* (D. Raven, W. Krohn, & R. S. Cohen (Eds.), Boston studies in the philosophy of science, Vol. 200). Dordrecht: Kluwer Academic.

Zusne, L., & Jones, W. H. (1982). *Anomalistic psychology*. Hillsdale: Lawrence Erlbaum.

Part III
Modern Technology Shapes Modern Science

Chapter 6
Iteration Unleashed. Computer Technology in Science

Johannes Lenhard

Abstract Does computer technology play a philosophically relevant role in science? The answer to this question is explored by focusing on the conception of mathematical modeling, how this conception is modified in computational modeling, and how this change is related to computer technology. The main claim states that computational modeling is geared towards iterative procedures which replace complicated or even intractable integrations. This shift is not a mere technicality, but presents a major conceptual transformation of modeling. At the same time, it is argued, the form and function of iterative procedures are dependent on the available computer technology.

A number of different cases, among them the Schrödinger equation in quantum chemistry and Markov chain Monte Carlo methods, are discussed that span iterative strategies in pre-computer, mainframe, and desktop-computer time. In methodological respect, bracketed iterations are discerned from exploratory iterations. Opacity and agnosticism are discussed as epistemic ramifications of the shift towards iterations.

1 Introduction

Computer technology is in widespread use in a great variety of contexts – scientific ones as well as non-scientific ones. Many people, asked to point out the signature technology of the recent decades, would single out computer technology – what makes it a prime example when one investigates the impact and importance of technology for science.

However, the term computer technology can be used in two senses. One is the technology of the computer as a machine, its hardware, like input and output devices, the integrated circuitry, or its principal design architecture etc. The other sense is technology that crucially depends on or entails computer technology in the first sense. Examples reach from the enormous Large Hadron Collider in particle

---

J. Lenhard (✉)
Philosophy Department, Bielefeld University, Pb. 100131, 33501 Bielefeld, Germany
e-mail: johannes.lenhard@uni-bielefeld.de

physics over computer tomography in medicine to the tiny dimensions of DNA arrays. The sheer range of different technologies that rely on computer technology testifies its relevance. The present chapter will mainly deal with computer technology in this second sense, though it will take into account that the second sense is not independent of the first one.

The key question is: Granted the importance of computer technology what is its impact on science? This question receives its significance when compared to an influential view, one could call it the standard account, that acknowledges the *practical* importance of computer technology, but more or less denies that it has a *philosophically relevant* impact on science. It holds that computer technology is a versatile and powerful tool of science, a means to accelerate and to amplify research and development, and also to boost all sorts of applications. According to this account, however, science would not have changed much in essence. Basically, computers do carry out computations and store and handle data. Both activities have been present in modern science all the time, hence do not strictly require computer technology. What computers do, so the standard account, is merely to speed-up data-handling and computation to astonishing degrees. Such a viewpoint denies a *philosophically* relevant role of computer technology in science.[1]

The present chapter will argue that this standard view is misleading, because computer technology does not present merely a new costume for basic mathematical calculation, rather it changes the very conception of mathematical modeling. Hence it changes the entire game. Consider the impact of the printing press. It would be grossly misconstrued if one would take it as a mere acceleration of writing by hand. What follows in this chapter is the attempt to convince you that this analogy, though somewhat grandiose, is an apt one.

Basically, computer technology leads to the automation of algorithmic procedures in a fairly wide sense. In his analysis of industrial economy, Karl Marx has pointed out that the key driver of the industrial revolution was not so much the power supply by the steam engine, but rather automation of artisan work, like the Spinning Jenny had achieved. Computer technology seems to hold a comparable position insofar as it automates formally or mathematically described procedures. However, such a comparison has to be based on firmer ground. This chapter describes and analyzes the impact of computer technology. What are its characteristics? What are its ramifications concerning the methodology and epistemology of science?

Section 2 takes "Representing and Intervening" as a starting point, the two major activities that Ian Hacking (1983) discerned in science. Hacking, among others, did much to revalue the intervention part and to confirm that both parts are philosophically on a par. It is an uncontroversial observation that most interventions into the material world, and experiments in particular, rely on technologies. Thus technology participates in the status of interventions. The significance of technology

---

[1] For instance, computing technology has not received an entry in the rich compilation of Meijers (2009), whereas diverse chapters discuss relevant examples.

for science has surely reached a new level due to computer technology. However, I want to put forward a stronger claim that holds that computer technology is interweaving representation and intervention and hence calling into question the distinction between the two.

Mathematical representation will serve as an instance. The function and form of mathematization changes with computing technology and this change is of utmost importance for the philosophy of science. This claim will be supported by an analysis of computational models and of the ways such models are constructed and implemented. In particular, computational modeling presents a fundamental transformation of mathematical modeling. This transformation can be characterized by the methodological notions of iteration and exploration, and by the epistemological notions of opacity and agnosticism. This transformation affects the perceived structure of the scientific endeavor.

The thesis will be detailed by discussing iteration, exploration (in Sect. 3), opacity, and agnosticism (Sect. 4) as major features of modeling under the conditions of computer technology. The upshot of the analysis will be that computer technologies and the ways they are employed in producing data, identifying phenomena, analyzing them, and making predictions, are mutually re-enforcing. Computer technology thus canalizes the trajectories of science, not in the sense of determining them, rather making some of them attractive. Section 5 will address the significance of infrastructure for how science is organized. The concluding section will indicate potential lessons one might draw for the study of the sciences.

2 Representing and Intervening

In his classic introduction to philosophy of science, Ian Hacking discerns two major scientific activities, those of representing and intervening (1983). We will use this differentiation as a starting point to assess the significance of computer technology.

One of Hacking's main points is to show that the (then) current tendency in philosophy of science, namely to see the representation part as the philosophically more relevant part of science, is an unwarranted prejudice. Both parts – representation and intervention – play their indispensable role; Hacking's famous slogan for instrumental realism – "If one can spray them, they exist" – may suffice as an example where he stresses the relevance of the intervention side. Of course, Hacking was by far not the only philosopher of science who pointed that out. The whole movement of so-called 'new experimentalism' more or less gives the activity of intervening center stage in science. It is uncontroversial that technology plays a major part in interventions, be it on the scale of laboratory experiments or that of engineering. Hence the significance of technology for science will in general be positively correlated with the valuation of interventions.

In his book Hacking forcefully grants intervention its place, but a place still somewhat derived from representation. The overarching view is rather one of "homo depictor" than the "homo faber".[2] The assessment of computer technology will lead us to place representation and intervention on a par, indeed to see both as deeply intertwined aspects of one activity. The following premise is surely justified: The significance of technology for science has reached a new level due to computer technology, because this type of technology is so widespread and versatile. So wherever technology comes in, say for example in a measurement apparatus, it is not unlikely that some computer technology is included. Admittedly, this claim leaves the question open what the characteristics of computer technology are. My point is not merely that this technology amplifies the share of the intervention part in science, rather computer technology questions the boundary between representing and intervening.

Representation has to do with identifying and displaying structures, and with creating a picture that helps to understand how phenomena come about and which factors contribute in what way to their dynamics. Computer technology, or more precisely, the use of computer technology in science, undermines this viewpoint. Computer models often do not represent what is out there, but play an essential role in transforming the subject matter and in constituting the phenomena. Sociologist of science Donald MacKenzie, for instance, supports a performativity thesis when he argues that computer models function in financial economics as: "An Engine, Not a Camera" (2006).

A classic topic in philosophy of science is that observations are always influenced by theory ('theory-ladenness'). Scholars like Paul Edwards rightly add that many data are 'model-laden' – his example are highly processed satellite data about the earth's climate (Edwards 2010). The present chapter argues that a similar view is valid in computational modeling more generally. The methodological and epistemological characteristics of computational modeling lead to an interweaving of representation and intervention.

3 Computational Modeling

Mathematical modeling has played an influential role in many modern sciences. A main impact of computer technology on science, or on scientific practice, was (and is) that it contributed to the birth of a new type of mathematical modeling, namely computational modeling. This new type is dependent on computational power and thus is tied to computer technology. There are further dependencies of technological nature, as we will see, insofar easy access to the machines, networked infrastructure, and the form of input and output interfaces matter.

---

[2]See Hacking 1983, 132. I would like to refer to Alfred Nordmann 2012 who argues about the relationship of representing in intervening in Hacking and about the broader significance for the philosophy of science and technology.

Computational modeling nowadays is employed in a great variety of sciences. It is quite obvious that the scope of mathematical modeling (in the dressing of computational modeling) has been enlarged to an astonishing degree. This observation urges us to ask how we can philosophically characterize computational modeling. There is a methodological and an epistemological part of the answer. Both parts point to the key role that technology plays in computational modeling.

3.1 Iteration

A preeminent feature of computational models is that they employ iterative strategies. Let us take quantum chemistry as an example and start with a look at the British mathematician and physicist Douglas R. Hartree (1897–1958). He was ahead of his time when he conceived computer technology and mathematical methods as twins. He combined great mathematical skills with a passion for tinkering and automating calculation procedures. He is mostly remembered for the anecdote, though one that actually happened, that he built a working copy of Vannevar Bush's Differential Analyzer out of Meccano parts that he diverted from his children's toy inventory. But he deserves a fuller appreciation in the context of our investigation. Hartree was not only an expert in computing technology in general, but was a very early and ardent follower of digital machines in particular. He conducted pioneering work with the ENIAC and the expertise he gained there was essential input for the development of the EDSAC at Cambridge, UK.

Early on, Hartree realized from his experiences with the ENIAC and its general-purpose programmability that digital computing would not merely make computation faster, but would demand to adapt mathematical modeling to the technology: "One thing which I learnt from practical experience of handling the ENIAC is the importance of taking what I might call a 'machine's-eye view' of the problems as furnished to it; and this is not so easy as it sounds." (1984 <1947>, 22) These problems covered technical issues like compiling a computer program that todays compilers and software languages solve automatically. The most important problem, however, was and is how mathematical problems should be formulated adequately. Although it is obvious that such formulations somehow have to reflect the technology available, it is not straightforward what that means for mathematical or computational modeling. High-speed machines were going to change methods and problems together, as Hartree expressed: "It is necessary not only to design machines for the mathematics, but also to develop a new mathematics for the machines." (1949, 115)

Hartree figured that iterative methods would suit particularly well to high-speed machines. Shockwaves, for instance, originate as singularities in solutions of a non-linear partial differential equation (PDE). If one replaces the PDE by a finite difference version, Hartree reasoned, one would need (under particular conditions) 200,000 multiplications to calculate an approximation to a practically useful degree. This condition had made such procedures impractical for extant devices. With the electronic digital computer, however, such iterative procedures had become a viable option and even a favorable way to go as iteration is exactly what the electronic digital computer is good at. Hartree foresaw a canalizing effect – mathematical modeling would, and should, develop in a way that adapts to the strengths and weaknesses of the computer's capability (1984 <1947>, 31).

Hartree was especially prepared to acknowledge the prospects of iterative methods because he had pioneered them in a half-automated setting even prior to the digital computer. In general, iterative processes are very suitable for mechanization – repeating similar operations – and Hartree was inventive to design an iterative procedure that navigated between technological feasibility and modeling accuracy, drawing on mechanized and graphical integration methods. This procedure was the starting point for the now prominent Hartree-Fock method in quantum chemistry. It is worth to view at this method because it displays an iterative strategy in the context of a computing technology prior to the electronic computer, even if it became popular with the digital computer (for more historical detail as well as more quantum chemical context, cf. Park 2009).

Hartree's work in quantum chemistry, or chemical physics, can be regarded as important early outsider contribution. The key problem of quantum chemistry is to solve the Schrödinger equation which contains – in principle, at least – the information about the electronic structure of atoms and molecules. This equation, alas, is an epitome of complexity and analytical solutions are practically impossible. Hartree conceived the challenge as one of creating a different mathematical strategy, not oriented at analytical solution, but numerical adequacy. For him, extracting numerically adequate values out of the Schrödinger equation was a challenge to computational modeling. His strategy was to jointly develop a model and a procedure (including technology) that together would be practically operational. He devised and designed also a procedure (including graphical integration steps) that could be iterated semi-mechanically and that was fitted to a model. In fact, it was a mutual fit: It was a key conceptual step to consider procedure and model as mutually dependent on each other and to develop both jointly.

Let me briefly consider the iterative character. The Schrödinger equation entails an elevated level of computational complexity, because each electron interacts with all others, so that the values for electron potentials cannot be calculated independently one after the other. Hartree's approach placed a numerical handle on this problem by constructing a different kind of iteration in the following way.

One starts to calculate the value of the potential of one electron (counterfactually) assuming all others fixed as a given (ad hoc) field. In the next step, the first electron is assigned the value calculated in the first step and the second electron is regarded as variable, all others as fixed. Then the numerical procedure is iterated. The first series is completed when each electron has been singled out for one step. Then the whole series of iterations gets iterated, i.e. one starts anew a second round of iterations, now taking the results of the first round as initial conditions. At the end of the second series of iterations, the values of the potential of all electrons have been re-adapted. If they differ from the first series, the whole procedure is iterated again. And so on, until the values do not change anymore between two series of iterations.

This iterative procedure is now known as self-consistent field (SCF) approach. It creates a balance through iteration. Ingeniously, one has ignored different parts of the interaction potential in each step to find values that are mutually consistent, hoping that errors then cancel out. This balance substitutes the analytical solution and emerges through a co-ordination of technology and modeling.

One can criticize Hartree's approach as not principled (enough) and too artificial, as merely oriented at numerical and mechanical feasibility. For Hartree, numerical virtues outweighed the gaps in the theoretical justification of the model. Hartree's method, or more precisely its refined descendant, the Hartree-Fock method, got widely accepted when computers became a crucial instrument in quantum chemistry and when it turned out that the predictions obtained by this method were deemed good enough.

The SCF procedure does away with the interdependence of electrons that is the main obstacle for computational solution strategies. Thereby it deliberately ignores the factually existing mutual interdependence. From our point of view, computational modeling inevitably does some harm to the in-principle rigor of quantum – or any fundamental – theory. However, any modeling attempt has to do this at one or the other spot. This perspective ties in with the recent debate in philosophy of science about the role of models and especially about "models as autonomous mediators" (Morrison 1999). Computer models are not defined and specified solely in terms of some underlying theory – although the latter might be very important, as in Hartree's case. Rather model and computer technology are interwoven. Hartree's point was that he took numerical feasibility as a guiding criterion for model-and-instrument together. His major contribution in philosophical respect, I would like to argue, is not his mastering the challenge of numerical feasibility, but that he approached model and instrument as a compound.

Thus, we have seen the proposed canalizing effect at work. The particular example of Hartree was included because it started from a strong fundamental theory with a clear mathematical formalization, i.e. the Schrödinger equation, so that it is a 'hard' case for showing the relevance of computer technology. One might ask: Can't that interdependence be taken into account later – by way of error corrections? Yes, it can. But this addition then has to build on and improve the SCF-based result, i.e. it constitutes a modeling step that itself is based on the performance of a numerical procedure. Consequently, this correction would have to follow a more exploratory than principled reasoning. Today, a family of numerical strategies called post-Hartree-Fock methods follow this path. They came up much later however, when supplementing the iterative strategy by an exploratory component was a good idea. Exploration is related to technology, too.

3.2 Grappling with Exploration

Let me explain why I take the difference between the iterative and the exploratory component as important. Iteration is a straightforward operation and iterative strategies are long known, they are for instance parts of the famous algorithms of Newton or Gauss. Computer technology can play out its strengths where modeling involves procedures that require so many iterative steps that they are infeasible without this technology. Additionally, there is also a basic difference in iterative strategies. Hartree's SCF method is *bracketed* in the sense that there is one command to conduct iterations until a certain state has been achieved which then is taken as result. Iterative strategies become *exploratory* when the procedures get interrupted and changed in light of the preliminary results. That is, one is not so much testing one single model, but rather testing and adapting a whole class of models in an exploratory manner. Therefore, feedback is an essential element in the modeling process. It is greatly enhanced by visual output, and it presupposes more or less easy accessibility so that researchers can afford to spend most of the time with preliminary versions of models.

My premise is: The exploratory add-on to iteration is a typical feature of smaller, i.e. highly available and cheap machines. Hence the exploratory-plus-iterative character of computational modeling is a feature of new computer technology since around 1990 that made these features available.[3] Let us approach the point of exploratory modeling strategies from an example where computer technology was not inviting. The next episode provides an illustration of how the digital computer opened up iterative strategies in quantum chemistry but at the same time also set limits regarding exploration.

Right from its invention in the late 1940s, the digital computer was related to the 'big science' of nuclear physics and the military. Peter Galison (1996), for instance, gives a vivid account of how Ulam, von Neumann, and others reasoned about computational strategies that would become tractable with a machine able to do high-speed iteration of elementary mathematical operations that otherwise would have demanded too high a number of human computers.

Other than physics, quantum chemistry was not 'big science' and to get access to computers was difficult in the 1950s. The case of Bernard Ransil who conducted the first so-called *ab initio* calculation might serve as an illustration. In 1956, Ransil worked at the University of Chicago in the quantum chemistry group of Robert Mulliken and Clemens Roothaan. At that time quantum chemical calculations normally followed a 'semi-empirical' approach, i.e. they inserted empirically determined values to simplify overly complex and difficult computations. For example, one would put in the value of a certain energy that is mathematically expressed in terms of an integral, if the value is known from experiment, but (too) hard to obtain by integration. Already very simple diatomic cases required impractically long computation times with extant mechanical devices – if one chose to refrain from the semi-empirical strategy.

Ransil's task was to design and implement the "first computer program to generate diatomic wavefunctions" (Bolcer and Hermann 1994, 8) without recourse to empirical values. This program was written in machine language for a UNIVAC

---

[3] This thesis is put into the context of a "culture of prediction" by Johnson and Lenhard (2011).

(Remington-Rand 1103) computer that was not in possession of the university, but of the military and was located at Wright Field Air Force Base in Dayton, Ohio. Mulliken had contracted excess computing time from the military. That meant a particularly inconvenient working arrangement: Ransil had to prepare the set of commands of his program, then travel from Chicago to Ohio with a stack of prepared punch cards, and work over night or the weekend with the UNIVAC (for more color to the story, see Mulliken 1989). The modification of the program was extremely tedious by today's standards due to the working conditions and also due to the fact that machine language programs will regularly require entirely new programming when the model is modified.

Nevertheless, at the end of the day, or many days, or rather many nights, the program ran. That gave reason for Mulliken and Roothaan to announce "Broken Bottlenecks and the Future of Molecular Quantum Mechanics", so the title of their (1959) paper. The broken bottleneck to which they refer was the complete automation of computing which made computational strategies feasible that avoided to bring in empirically determined values. They report about Ransil's machine program: "The importance of such a machine program is illustrated by the fact that the entire set of calculations on the N_2 molecule which took Scherr (with the help of two assistants) about a year, can now be repeated in 35 min ..." (Mulliken and Roothaan 1959, 396).

The speed of computation definitely mattered and speed was basically a matter of technology. At the same time, it was a demanding process to arrange all conditions so that one could profit from the speed. On the one hand, computer technology unleashed iterative modeling strategies. On the other hand, it erected also limiting conditions due to cost, availability, programming language, and (missing) feedback interfaces. In effect, exploratory work was hardly possible, because it would have demanded an infrastructure where going back-and-forth between testing the performance and modifying the model would be a practical option. This was clearly not the case at Ransil's time and, more general, exploration was hard during the time of relatively expensive and centrally maintained mainframe computers. They canalized modeling toward bracketed rather than exploratory iteration.

3.3 *From Bracketed to Exploratory Iteration*

Quantum chemistry is arguably only one case among many, picked out here because it covers both pre-computer and computer era.[4] Let us now look at a different case, not a particular scientific field, but a technique that is used in a wide array of sciences, namely the Monte Carlo (MC) method. It is discussed, for instance, by Peter Galison (1996) who wants to make the case for a "Tertium Quid", locating MC

---

[4]Lenhard (2014) gives a more detailed account of the development of computational quantum chemistry.

simulation alongside experiment and theory. The issue of simulation is discussed in Chap. 7 of this volume; in the following we focus on the iterative nature of MC methods.

Consider a simplistic example of how MC works: Calculating the area of a figure is often difficult because integration depends on the algebraic form of the 'boundary'. MC integration provides an elegant way out: Put the figure in a rectangular frame of known size and create – with the help of a (pseudo-)random number generator – points with random positions inside the frame. Some points will fall into the figure, some will lie inside the frame but outside the figure. If one iterates this procedure very often, i.e. if one creates very many random points, the mathematical law of large numbers will be at work and guarantee that the 'empirical' fraction of points in the figure relative to all points will converge to the relative size of the figure in the frame. Thus, the basic recipe of MC is to replace *analytic integration* by a *probabilistic* approach with massive *iteration* – a method only feasible based on computer technology.

Monte Carlo methods rest on a fundamental conceptual shift while on the algorithmic level, it works with brute force iteration. A well-known downside is the rather slow convergence rate, i.e. one needs often impractically long runs to reach the desired precision. Such methods have received great sophistication in the form of so-called Markov chain Monte Carlo (MCMC) methods. These methods have seen an enormous uptake and have become the standard approach to tackle complicated integrations in many branches of the sciences.

They combine MC with the theory of Markov chains, so let me first add a few words about Markov chains. Such chains are random processes that can be described as movements through a space according to probabilistic rules. Intuitively, think of a tourist's random walk through the streets of a town, choosing on each crossing between all streets with specified probabilities. After a couple of decisions, an observer can still guess where roughly the tourist started, but after many steps this will be less so. The basic mathematical theorem[5] posits that such a chain will converge to its unique equilibrium distribution, no matter where it started. A most astonishing fact to experts in the field is that convergence to equilibrium regularly happens very quickly! In typical cases, the speed of convergence is a fact observed from computer runs, not a general fact derived from Markov chain theory.

Markov chain Monte Carlo methods make use of this fact to simulate the equilibrium distribution. Again, the algorithm is easy to implement. Let the Markov chain move for a while, report where it is, then start over anew. If one repeats that often, the cumulated reported results will present a Monte Carlo picture of the equilibrium distribution. The point is that the convergence rate of this iterative procedure is much higher than for regular Monte Carlo. Thus it is easy to simulate the equilibrium distribution for a given Markov chain. The trick of MCMC is to start

---

[5]Matters are greatly simplified in our discussion. Only the discrete case is considered and questions of how the space is defined or which technical conditions have to be satisfied are ignored as they are not important for the illustrative task.

with a complicated object, like a high-dimensional integral, and then to construct a Markov chain that has this object as equilibrium distribution. This construction is often not difficult – mainly because one has the license to define the Markov process in a suitable space, like the space of configurations of the Ising model (discussed in a moment). The basic conceptual point of MCMC is the same as with the simpler Monte Carlo, i.e.

(i) complicated or analytically intractable integrals are transformed into the idiom of repeated (random) processes and thereby
(ii) the problem of integration is transformed into one of iteration.

MCMC has roots practically as old as the Monte Carlo method itself; it goes back to Metropolis et al. (1953) and was later systematized in the Metropolis-Hastings-algorithm (Hastings 1970). Although the method was specific enough to be called an 'algorithm', MCMC acquired great popularity only since around 1990. This more recent dynamics is based on the interplay of computer technology and modeling that changed MCMC from a somewhat curious invention into a method of extremely wide use. The key, or so will be argued, is the step from bracketed to exploratory iteration.

Let us illustrate this claim by a standard example, the Ising model of thermodynamics. Consider a two-dimensional grid of cells; each cell has a spin (up or down) and interacts with its four neighbors via a tendency to take up the same spin as the neighbor. This behavior is implemented as a stochastic process. Roughly, in each time step the spin at any location takes on the same value as the spin at neighboring locations, but only with a certain probability. And the probability is higher the more neighbors already show a particular spin. Local interactions are easily described while the resulting global distributions are famously intractable, because the state of the neighbor of a given cell depends again on the states of its neighbors etc.[6] One spectacular proof of success of MCMC was the solution of the Ising model's riddle. MCMC transformed the intractable problem of determining the equilibrium distribution into a question that was solvable with a surprisingly moderate effort.[7] MCMC became a standard method to tackle complicated probability distributions and multi-dimensional integrals that formerly were deemed intractable.

The iterative part is clearly fundamental to MCMC, but what is the role of the exploratory component? This role comes to the fore when one addresses the question whether the Markov chain has actually reached the equilibrium. It is known from 'experience' that such chains often converge surprisingly fast, but there is no known general mathematical result stating how fast a given chain converges and when it has reached its equilibrium. Diaconis (2008) acknowledges the tremendous

---

[6]The problem is similar to the one of electron interaction in the Schrödinger equation.

[7]Persi Diaconis, a leading probability theorist, vividly describes how astonished he was when he first saw the MCMC solving this task (2008). R.I.G. Hughes (1999) gives a highly readable account of the Ising model in the context of simulation.

impact of MCMC on recent sciences and he aptly points out that the convergence question is an important and urgent problem.

Though there are no strict solutions to this problem, there exist a bunch of strategies to explore and inspect the behavior. The software AWTY (Nylander et al. 2008) that has been designed for the special case of MCMC methods in Bayesian phylogenetics provides an illustration. The acronym expands in "Are we there yet?", i.e.: has the chain under consideration already reached equilibrium? The program provides visualizations of the chain's behavior that shall enable the researchers to explore the chain's behavior and to judge whether the dynamics looks like one near equilibrium. In effect, exploration with the help of visualization substitutes a theoretical convergence result. Regarding technology, this substitution process, and exploratory strategies in general, does not demand great computing power, rather it makes use of small and easily available computers that make such exploratory procedures a practical option. This observation indicates that the upswing of MCMC and the availability of (relatively small) computers at the same time are more than a mere coincidence.

4 Epistemic Ramifications

A good way to characterize the epistemic ramifications of computing technology is again by contrast. Mathematical modeling operationalizes all sorts of relationships and makes them amenable to systematic manipulation. Notwithstanding the fact that mathematical calculations might become complicated and demanding, mathematical modeling was oriented at, or rather bound to, a transparent outline. Arguably mathematics even served as a paradigm of epistemic transparency. This goal resonates with the metaphysical vision of simple fundamental laws that are expressed in mathematical form. It might even be a main factor in producing this vision.

Of course, mundane questions of tractability influenced mathematical model building. Consider systems of partial differential equations (PDEs) that come up in many instances when mathematical models express global relationships with the rationale of the differential and integral calculus. We have discussed already the Schrödinger equation; another famous example are the Navier-Stokes-equations in hydrodynamics. Such systems of PDEs remain largely intractable and hence the engineering sciences preferred approaches that circumvent such systems. With computer technologies, however, this attitude has reversed. Systems of PDEs now look very attractive because they can be treated by standard software packages that convert the analytically intractable into iterative schemes tractable by the computer.

However, the shift to computer technology has important epistemic ramifications that basically run counter to the epistemic transparency connected to mathematical modeling. Iterative procedures are fruitful exactly because iterations do not simply arrive at a point foreseeable in the beginning. Computer models, at least the more complex ones, perform a myriad of simple steps – every single one perfectly conceivable and 'logical'. Over run time, however, the initial assumptions interact in an intricate way and normally this cumulative effect creates epistemic opacity in the sense that manifest features of the model's behavior often cannot be traced back to particular assumptions or input values. Hence, mathematical modeling – in the form of computational modeling – is moving away from its own former virtue of epistemic transparency![8]

We have seen in the case of the Ising model and the MCMC method that the solution is computed although a transparent understanding is not attained. We are at a loss for two reasons. First, the definition tells us everything about the behavior of the Markov chain in the next time step, but tell us little about the long-term behavior, so that we (or the computer) have to actually conduct the iterations. Second, the definition itself follows an instrumental reasoning: The chain is not derived from its significance in terms of the dynamics in question (Ising model), apart from the sole aspect of having the desired equilibrium distribution. Hence there are often no intuitive clues of how such chains evolve over time.

Let us contemplate a bit more on the notion of understanding and its relationship to transparency and opacity. True, the ideal of epistemic transparency seems to be hampered when computer technology is at work. But how strict a prerequisite is this ideal for understanding? The latter notion is a quite flexible one – what counts as intelligible has undergone changes in history as highlighted by Peter Dear (2006), among others. He discerns two branches of science – natural philosophy oriented at explanation and understanding on the one hand and, on the other hand, science as instrumentality that is oriented at intervention.

"Alongside science as natural philosophy, therefore, we have science as an operational, or instrumental, set of techniques used to do things: in short, science as a form of engineering, whether that engineering be mechanical, genetic, computational, or any other sort of practical intervention in the world." (Dear 2006, 2)

Dear diagnoses that the parts are normally taken to relate to each other in a circular way:

"Why are science's instrumental techniques effective? The usual answer is: by virtue of science's (true) natural philosophy. How is science's natural philosophy shown to be true, or at least likely? The answer: by virtue of science's (effective) instrumental capabilities. Such is the belief, amounting to an ideology, by which science is understood in modern culture. It is circular, but invisibly so." (2006, 6)

Hence, science belongs to two ideal types simultaneously. Modern science emerged as natural philosophy amalgamated with instrumentality. Does computer technology change the composition of this amalgam? Does it affect the very notion of understanding?

---

[8]Here, my argument takes up Paul Humphreys' account (2004, 2009) of simulation who points out that epistemic opacity is an important (if deplorable) aspect of simulation. I would like to maintain that opacity applies to the use of computer technology more generally. As this topic is also discussed in Chap. 7, I can be brief here and just point out the general significance in connection with computer technology.

Dear stresses that there are no timeless, ahistorical criteria for determining what will count as satisfactory to the understanding. Assertions of intelligibility can be understood only in the particular cultural settings that produce them. Intelligibility, for him, is ultimately an irreducible category (2006, 14). This is the clue to my claim about the broader epistemological significance of computing technology: basic epistemic notions are related to technology – in the sense that they will change in correspondence to changes in technology. In our case, computational modeling affects what is desired and accepted as understanding.

To a certain extent, it is the mathematical form of representations that enables scientists and engineers to draw conclusions and to make predictions. Traditionally, in the circular picture Dear shows us, understanding, explanation, and the capacity for intervention are correlated in a positive way, i.e. the more understanding and explanation, the more ability to predict and to manipulate – and vice versa. With computer technology and computational modeling, however, the ability to predict (and therefore to systematically intervene) are *negatively* correlated to epistemic transparency.

At least, the use of computing technology together with computational models in some cases does provide accurate results. And this might be seen as a sufficient criterion for understanding: If you want to prove that you understand how a problem can be solved – show the solution. From this pragmatic perspective, it is a minor aspect how 'transparent' your machinery is. Thus computer technology seems to fit to a pragmatic, intervention-oriented notion of understanding. The somewhat paradoxical diagnosis then is that this pragmatic understanding is rooted more in technology than in intellectual transparency.[9] If we accept this pragmatic notion then explanation would be decoupled from understanding, because explanation seems to demand more and is less open to a pragmatic account. At least this holds for most philosophical accounts of scientific explanation. I see it as an open question whether the pragmatic sense of understanding will prove to be a preliminary and deficient mode – or whether it will be accepted as the thing that computational science can achieve and that eventually will be adopted as a (non-deficient) notion of intelligibility.

Additionally, the epistemic impact of computer technology on science is fostered by methods that are data-driven and do largely ignore any theoretical structure, but are nevertheless effective in yielding predictions. In their paper on "Agnostic Science", Napoletani, Panza, and Struppa (2011) describe the "microarray paradigm", a methodological paradigm of data analysis. They "argue that the modus operandi of data analysis is implicitly based on the belief that if we have collected enough and sufficiently diverse data, we will be able to answer any relevant question concerning the phenomenon itself." (2011, 1)

A microarray, also known as DNA array, is a chip with thousands of manufactured short strands of DNA on it. If a sample is tested (washed over it), constituents of the sample bind to the chunks on the chip, depending on both the composition of the probe and the spatial distribution of the strands in the array.

---

[9]Cf. Lenhard 2009 for a more full-fledged argument about understanding and simulation.

Thus, the resulting pattern – created with the help of artificial coloring of the probe – somehow mirrors what the probe consists of. However, the exact shape of the patterns depends on complicated conditions that are hard to control, like the exact location and constituents of the single DNA strands. In sum, the resulting patterns contain a wealth of information, but at the same time a high degree of noise. Because of the sheer bulk of data, even a high level of noise leaves intact the chances to detect the signal, i.e. extract relevant information about the probe. Efron (2005) highlights a similar perspective when he sees the twenty-first century marked by data giganticism, and takes microarrays as paradigmatic. In this regard, DNA arrays are a typical example of a situation where new technological high-throughput devices deliver great amounts of data. These data, in turn, require the use of computer technology for analysis.

The point is that computer based data analysis might be able to detect signals and to make predictions, like: the patterns of the probe resemble patterns produced by tissue with a certain disease. In this way, data analysis can take advantage of the amount of data without specifying a model of the dynamics that produces the data. This is what Napoletani et al. call "agnosticism". This agnosticism refers to a characteristic of the mathematical methods: They work on resemblance of patterns but do not involve theoretical hypotheses or structural assumptions about how the patterns are produced. In this respect, the mathematical techniques of data analysis are indeed "agnostic". Consequently, the success of DNA arrays and other data-driven methods shows how computer technologies on the data side and on the analysis side work together in a way that features prediction and intervention.

I want to avoid a potential misunderstanding: It is not claimed that science, due to computer technology, can get rid of any structural understanding. Rather, a new problem occurs to integrate the combined results of computer technology-and-modeling with more traditional approaches. Consequently, Napoletani et al. take a careful stance in later passages of their text. They suggest that after having achieved predictive success in an "agnostic" manner, later steps, especially the incorporation into the body of scientific knowledge, may need a more structure-based understanding.

5 Infrastructure

In this section I want to very briefly address how computer technology, in particular networked infrastructure and software packages, affects the social organization of science. Of course, there is excellent scholarly work on 'big science' as it was organized in the Manhattan Project and later on. Not accidentally, computer technology is a core element in big science, see for instance Edwards (1996). Climate science provides a somewhat different type of example, also "big", and also making essential use of computer technology (cf. Edwards 2010). These cases will not be the issue here, however. Instead, I want to concentrate on small computers that have become part of everyday science culture, even of everyday culture.

Sociologist of science Sherry Turkle has studied how design and education in architecture and civil engineering have changed in the course of the proliferation of computer technology (Turkle 2009). Her study compares two timelines, around 1980 and around 2000, i.e. before and after the spread of small computers. By 1980 computer technology had been introduced, but many practitioners remained hesitant so that the technology was employed in a way that was crafted on the older tool-building tradition. However, by 2000, the situation had changed, so Turkle. Many users were working with these tools as a matter of course while persons from the tool-building tradition that could repair their instruments or check the code weren't available anymore. That is, developers and users of software had started to build different social groups.

Turkle reports disagreement among architects as well as civil engineers whether students should learn programming (2009, 19) – does one need to understand the instruments one uses? On the one side, the use of software packages, like for computer assisted design, offered advantages in terms of which tasks could be fulfilled by persons without long experience in the field. On the other side, there were concerns about growing dependency on tools that essentially had become opaque.

Moreover, also the possibility of a new sort of understanding comes up in Turkle's study, one that is based on the *explorative* mode of modeling: At the 1980 timeline, design was thought to follow or elaborate on a fundamental plan. This changed, as Turkle reports (2009, 23), because computer programs allowed to play with preliminary designs. The exploratory mode can be recognized also in software tools that are adapted to different environments. To grab code and to customize it had become usual practice by 2000. These somewhat unprincipled procedures are reflected in a relatively low status of any single program. To use several programs and models and to compare their results is widespread practice that is greatly enhanced by infrastructure of networked computers.

Without doubt, these developments have serious drawbacks. There is a trade off between pragmatically motivated exploration and theoretically founded certainty. It is not yet clear, I would like to argue, to which extent scientists can influence, are able to choose, or rather have to accept how weights are assigned in this trade.

6 Conclusion

Let us take stock. We have investigated several aspects of how computer technology and conceptions of computational modeling are interrelated. In the form of computational modeling, mathematical modeling has undergone a fundamental transformation, characterized by the features of iteration, exploration, opacity, and agnosticism. This transformation affects the perceived structure of the scientific endeavor.

We only briefly discussed how computer technology is involved in the production of data. Admittedly, this constitutes a highly relevant matter. Data are – contrary to their etymological roots – not primarily 'given', but often construed by computer technology and computational models. Think of a CT scan, an image of a scanning tunnel microscope, a visualization of particle collisions at the LHC (Cern), or a map displaying rain clouds based on satellite data. All of them show highly processed data. Normally such data can only be produced and handled with the aid of computer technology. Data-driven methods have become a slogan in many branches of science. We have touched upon this issue during the discussion of agnosticism and the "DNA-array paradigm". The interplay of data production and analysis, made possible and mediated by computer technology, fosters our claim that technology and modeling are interwoven.

Finally, I would like to point out two issues that pose open questions. First, remind the metaphor of computer technology as the Spinning Jenny of computation. I take it as an open question whether science is inherently industrial or artisan, i.e. to which extent computer technology will change the fundamentals of scientific knowledge production. Does the exploratory mode yield only successes of a transient nature? Does the use of computer technology require to re-assess the notion of understanding in the way indicated in the preceding investigation? These questions have profound implications for how we understand science and our culture. Any answer will have to grant computer technology a central place in the analysis.

The second issue is a methodological one. If a study wants to elaborate on the previous argumentation, or to disprove it in some controversial points, it will profit when it combines philosophical, historical, and sociological aspects. I am convinced that attempts to elucidate the relationship between science and computer technology call for a re-assessment of how we study science and technology. There exist ongoing broader movements into that direction, like the recent 'practice turn' in philosophy of science, or integrated programs like "&HPS".[10] In my opinion, the explicit inclusion of technology and the technological sciences will be among the success conditions for this kind of study.

References

Bolcer, J. D., & Hermann, R. B. (1994). Chapter 1: The development of computational chemistry in the United States. In K. B. Lipkowitz & D. B. Boyd (Eds.), *Reviews in computational chemistry* (Vol. 5). New York: VCH Publishers.
Dear, P. (2006). *The intelligibility of nature. How science makes sense of the world*. Chicago: The University of Chicago Press.
Diaconis, P. (2008). The Markov chain Monte Carlo revolution. *Bulletin of the American Mathematical Society, 46*, 179–205.
Edwards, P. N. (1996). *The closed world. Computers and the politics of discourse in Cold War America*. Cambridge, MA: MIT Press.

---

[10] Among them also Nordmann (2012) who formulates a program for history, philosophy, and sociology of the technological sciences.

Edwards, P. N. (2010). *A vast machine. Computer models, climate data, and the politics of global warming*. Cambridge, MA: MIT Press.

Efron, B. (2005). Bayesians, frequentists, and scientists. *Journal of the American Statistical Association, 100*(469), 1–5.

Galison, P. (1996). Computer simulations and the trading zone. In P. Galison & D. J. Stump (Eds.), *The disunity of science: Boundaries, contexts, and power* (pp. 118–157). Stanford: Stanford University Press.

Hacking, I. (1983). *Representing and intervening. Introductory topics in the philosophy of natural science*. Cambridge: Cambridge University Press.

Hartree, D. R. (1949). *Calculating instruments and machines* (photographical reprint). Urbana: The University of Illinois Press.

Hartree, D. R. (1984). *Calculating machines. Recent and prospective developments and their impact on mathematical physics*. Cambridge, MA/London: MIT Press (Photographical reprint of Hartree's inaugural lecture 1947.)

Hastings, W. K. (1970). Monte Carlo sampling methods using Markov chains and their applications. *Biometrika, 57*(1), 97–109.

Hughes, R. I. G. (1999). The Ising model, computer simulation, and universal physics. In M. Morgan & M. Morrison (Eds.), *Models as mediators* (pp. 97–145). Cambridge: Cambridge University Press.

Humphreys, P. (2004). *Extending ourselves. Computational science, empiricism, and scientific method*. New York: Oxford University Press.

Humphreys, P. (2009). The philosophical novelty of computer simulation. *Synthese, 169*(3), 615–626.

Johnson, A., & Lenhard, J. (2011). Towards a culture of prediction: Computational modeling in the era of desktop computing. In A. Nordmann, H. Radder, & G. Schiemann (Eds.), *Science transformed? Debating claims of an epochal break* (pp. 189–199). Pittsburgh: University of Pittsburgh Press.

Lenhard, J. (2009). The Great Deluge: Simulation modeling and scientific understanding. In H. de Regt, S. Leonelli, & K. Eigner (Eds.), *Scientific understanding. Philosophical perspectives* (pp. 169–186). Pittsburgh: University of Pittsburgh Press.

Lenhard, J. (2014). Disciplines, models, and computers. The path to computational quantum chemistry. *Studies in History and Philosophy of Science Part A, 48*, 89–96.

MacKenzie, D. (2006). *An engine, not a camera. How financial models shape markets*. Cambridge, MA/London: MIT Press.

Meijers, A. (2009). *Philosophy of technology and engineering sciences* (Handbook of the philosophy of science, Vol. 9). Amsterdam: Elsevier.

Metropolis, N., Rosenbluth, M. N., Rosenbluth, A. H., & Teller, E. (1953). Equations of state calculations by fast computing machines. *Journal of Chemical Physics, 21*(6), 1087–1092.

Morrison, M. (1999). Models as autonomous agents. In M. Morgan & M. Morrison (Eds.), *Models as mediators* (pp. 38–65). Cambridge: Cambridge University Press.

Mulliken, R. S. (1989). *Life of a scientist*. New York: Springer.

Mulliken, R. S., & Roothaan, C. C. J. (1959). Broken bottlenecks and the future of molecular quantum mechanics. *Proceedings National Academy of Sciences, 45*, 394–398.

Napoletani, D., Panza, M., & Struppa, D. (2011). Agnostic science. Towards a philosophy of data analysis. *Foundations of Science, 16*(1), 1–20.

Nordmann, A. (2012). Im Blickwinkel der Technik. Neue Verhältnisse von Wissenschaftstheorie und Wissenschaftsgeschichte. *Berichte Wissenschaftsgeschichte, 35*, 200–216.

Nylander, J. A., Wilgenbusch, J. C., Warren, D. L., & Swofford, D. L. (2008). AWTY (are we there yet?): A system for graphical exploration of MCMC convergence in Bayesian phylogenetics. *Bioinformatics, 24*(4), 581–583.

Park, B. S. (2009). Between accuracy and manageability: Computational imperatives in quantum chemistry. *Historical Studies in the Natural Sciences, 39*(1), 32–62.

Turkle, S. (2009). *Simulation and its discontents*. Cambridge, MA/London: MIT Press.

Chapter 7
Computer Simulations: A New Mode of Scientific Inquiry?

Stéphanie Ruphy

Abstract Computer simulations are everywhere in science today, thanks to ever increasing computer power. By discussing similarities and differences with experimentation and theorizing, the two traditional pillars of scientific activities, this paper will investigate what exactly is specific and new about them. From an ontological point of view, where do simulations lie on this traditional theory-experiment map? Do simulations also produce measurements? How are the results of a simulation deem reliable? In light of these epistemological discussions, the paper will offer a requalification of the type of knowledge produced by simulation enterprises, emphasizing its modal character: simulations do produce useful knowledge about our world to the extent that they tell us what could be or could have been the case, if not knowledge about what is or was actually the case. The paper will also investigate to what extent technological progress in computer power, by promoting the building of increasingly detailed simulations of real-world phenomena, shapes the very aims of science.

1 Introduction

In 2013, two projects were selected by the European Commission as "Flagships" projects, receiving each a huge amount of funds (about one billion euros over 10 years). It is telling that one of these two top-priority projects, the Human Brain Project, aims at digitally *simulating* the behaviour of the brain. Computer simulations have not only become ubiquitous in the sciences, both natural and social, they are also more and more becoming ends in themselves, putting theorizing and experimenting, the two traditional pillars of scientific activities, into the background. This major addition to the range of scientific activities is in a straightforward sense directly linked to technological advances: the various epistemic roles fulfilled by computer simulations are inseparable from the technology used to perform it, to wit, the digital computer. Asking to what extent technology (in that case

---

S. Ruphy (✉)
Laboratoire PPL, Université Grenoble Alpes, F-38000 Grenoble, France
e-mail: stephanie.ruphy@upmf-grenoble.fr

ever increasing computing power) shapes science requires assessing the novelty and the epistemological specificities of this kind of scientific activities. Sure enough, computer simulations are everywhere today in science – there is hardly a phenomenon that has not been simulated mathematically, from the formation of the moon to the emergence of monogamy in the course of evolution of primates, from the folding of proteins to economic growth and the disintegration of the Higgs boson, but what exactly is specific and new about them?

There are two main levels of assertions about their novelty in the current philosophical landscape. A first kind of assertions concerns the extent to which computer simulations constitute a genuine addition to the toolbox of science. On a second level, the discussion is about the consequences of this addition for philosophy of science, the question being whether or not computer simulations call for a new epistemology that would be distinct from traditional considerations centered on theory, models and experiments.

Given the topic of this volume and the direct link between technological progress made in computational power and simulating capacities, I will be mainly interested in this paper in the first kind of assertions, the ones that state the significant novelty of computer simulations as a scientific practice.[1] Here's a sample of those claims, coming both from philosophers of science and scientists. For the philosopher Ronald Giere for instance, the novelty is quite radical: "[...] computer simulation is a qualitatively new phenomenon in the practice of science. It is the major methodological advance in at least a generation. I would go so far as saying it is changing and will continue to change the practice not just of experimentation but of science as a whole" (2009, 59). Paul Humphreys, also a philosopher of science, goes even one step further by talking about revolution: "[...] computer modelling and simulation [...] have introduced a distinctively new, even *revolutionary*, set of methods in science" (2004, 57. My italics). On the scientific side, the tone is no less dramatic as for instance in a report a few years ago to the US National Academy of Sciences: "[But] it is only over the last several years that scientific computation has reached the point where it is on a par with laboratory experiments and mathematical theory as a tool for research in science and engineering. The computer literally is providing a *new window through which we can observe the natural world in exquisite detail.*" (J. Langer, as cited in Schweber and Wächter 2000, 586. My italics).

In their efforts to further qualify the novelty of computer simulations and the associated transformative change of scientific activities, philosophers of science have engaged into descriptive enterprises focusing on particular instances of simulation. Given the widespread taste of professional philosophers for accumulation of definitions and distinctions, as well as for fine-grained typologies, efforts have also been made to offer scientifically informed definitions of simulations (distinguishing them in particular from models), as well as typologies ordering the variety of

---

[1] See Humphreys (2009) and Reiss and Frigg (2009) for discussions of the second kind of assertions.

scientific enterprises coming under the banner of computer simulation, by typically classifying them according to the type of algorithm they employ ("Discretization" mathematical techniques, "Monte Carlo" methods, "Cellular automata" approaches, etc.). However interesting and useful these philosophical studies are, I won't talk much here of the various definitions and distinctions they propose, being more concerned by the challenging epistemological and ontological issues *common* to many kinds of simulations. And a widely-discussed first set of issues refers to the relationship between computer simulations and experimenting and theorizing. From an ontological point of view, where do simulations lie on this traditional theory-experiment map? Simulations are often described as "virtual" or "numerical" experiments. But what are the significant similarities or differences between computer simulations and experiments? Do simulations also produce measurements? Do they play similar epistemological roles *vis-à-vis* theory? Another set of challenging issues concerns the sanctioning of a computer simulation. How do computer simulations get their epistemic credentials, given that they do not simply inherit the epistemic credentials of their underlying theories (Winsberg 2013)? Is empirical adequacy a sure guide to the representational adequacy of a simulation, that is, to its capacity to deliver reliable knowledge on the components and processes at work in the real-world phenomenon whose behaviour it purports to mimic? As we shall see, this kind of issues are especially acute for what I will call *composite* computer simulations, developed to integrate as much detail of a given phenomenon as computing power allows. In light of these epistemological discussions, I will offer a requalification of the type of knowledge produced by simulation enterprises, emphasizing its *modal* character. And I will conclude with tentative remarks on the way ever increasing computing power, by promoting the building of fully detailed simulations of real-world phenomena, may progressively transform the very *aims* of science.

2 Hybrid Practice

A good starting point to discuss the similarities and differences between simulations on the one hand, and experiments and theories on the other, might be to ask scientists how they would describe their activities when they build and use simulations. Fortunately, some science studies scholars have done just that and I will draw here on Dowling's (1999) account based on 35 interviews with researchers in various disciplines ranging from physics and chemistry to meteorology, physiology and artificial life. One of the most interesting, if not totally surprising lessons of Dowling's inquiry is that for its practitioners the status of this activity is often hybrid, combining aspects partaking of theoretical research and of experimental research. Simulations are commonly used to explore the behaviour of a set of equations, constituting a mathematical model of a given phenomenon. In that case, scientists often express the feeling that they are performing an experiment, by pointing out that many stages of their digital study are similar to traditional stages of an experimental work. They first set some initial conditions for the system, vary the values of parameters, and then observe how the system evolves. To that extent, as in a physical experiment, the scientist interacts with a system (the mathematical model), which sometimes may also behave in surprising ways. In other words, in both cases, scientists engage with a system whose behaviour they cannot totally anticipate, and that is precisely the point: to learn more about it by tinkering and interacting with it. In the case of a simulation, the unpredictability of the system is no mystery: it usually comes from the nature of the calculations involved (often dealing with non linear equations that cannot be solved analytically). Scientists sometimes talk about "the remoteness of the computer processes": they cannot fully be grasped by the researcher who "black-boxes" them while performing the simulation run (Dowling 1999, 266).

Mathematical manipulation of a theoretical model is not the only experimental dimension of a computer simulation. Producing data on aspects of a real-world system for which observations are very scarce, inexistent or costly to obtain is another widespread epistemic function of a computer simulation. To the extent that these simulated data are then often used to test various hypotheses, computer simulations share with experiments the role of providing evidence in support or against a piece of theoretical knowledge.

As for the similarities with theories, the point has been clearly, if somewhat simplistically, made by one of the physicists being interviewed: "Of course it's theory! It's not real!" (Dowling 1999, 265). In other words, when the issue of the relationship to reality is considered, that is, when simulations are taken as representations, the manipulation dimension of the simulation gives way to the conjectural nature it inherits from its theoretical building materials. So from the point of view of the practioners, computer simulations combine significant features of both theories and physical experiments, and that might explain why simulation practioners are sometimes less inclined than philosophers to describe computer simulation as a radically new way of finding out about the world. That might also explain why expressions such as "in silicon experiments", "numerical experiments", "virtual experiments" have become so common in the scientific discourse. But from an ontological point of view, to what extent exactly should these expressions be read literally?

Philosophers of science have further explored the similarities between computer simulations and physical experiments by asking three kinds of (related) questions. First, can one still talk of experiment in spite of the lack of physical interactions with a real-world system? Second, do simulations also work as measurement devices? Third, does the sanctioning of a computer simulation share features with the sanctioning of an experiment?

3 The 'Materiality' Debate

The first kind of questions is often referred to as the 'materiality' debate (see e.g. Parker (2009) for a critical overview of it).[2] This debate builds on the claim that the material causes at work in a numerical experiment are (obviously) of a different nature than the material causes at work in the real-world system (the target system) being investigated by the simulation, which is not the case with a physical experiment as explained by Guala:

> The difference lies in the kind of relationship existing between, on the one hand, an experiment and its target system, and on the other, a simulation and its target. In the former case, the correspondence holds at a "deep", "material" level, whereas in the latter, the similarity is admittedly only abstract and formal. [...] In a genuine experiment, the same material causes as those in the target system are at works; in a simulation, they are not, and the correspondence relation (of similarity or analogy) is purely formal in character (2005, 214–215).

This ontological difference emphasized by Guala has epistemic consequences. For Morgan (2005) for instance, an inference about a target system drawn from a simulation is less justified than an inference drawn from a physical experiment because in the former case, and not in the latter case, the two systems (the simulation/experimental system and the target system) are not made of the "same stuff". In other words, as Morgan puts it, "ontological equivalence provides epistemological power" (2005, 326). And computer simulations, if conceived as experiments, must be conceived as *non-material* experiments, on mathematical models rather than on real-world systems. This lack of materiality is precisely what Parker wants to challenge. Parker makes first a distinction between a computer simulation and a computer simulation study (2009, 488). A computer simulation is a "sequence of states undergone by a digital computer, with that sequence representing the sequence of states that some real or imagined system did, will or might undergo" (2009, 488). A computer simulation *study* is defined as "the broader activity that includes setting the state of the digital computer from which a simulation will evolve, triggering that evolution by starting the computer program that generates the simulation, and then collecting information regarding how various properties of the computing system [...] evolve in light of the earlier information" (2009, 488). Having defined an experiment as an "investigative activity involving intervention" (2009, 487), Parker then claims that computer simulation *studies* (and not computer *simulations*) do qualify as experiments: when performing a computer simulation study, the scientist does intervene on a *material* system, to wit, a programmed digital computer. So in this particular sense, concludes Parker, computer simulation studies are *material* experiments: 'materiality' is not an exclusive feature of traditional experiments that would distinguish them from computer studies.

For all that, acknowledging this kind of materiality for computer simulation studies does not directly bear on Morgan's epistemological contention. Recall that

---
[2] See also Barberousse et al. (2009) and Norton and Suppe (2001).

the epistemic advantage granted to traditional experiments follows from the fact that the system intervened on and the target system are made of "the same stuff" and not only from the fact that both systems are material systems. In the case of a computer simulation study, the system intervened on and the target system are both material but obviously they are not made of the same kind of "stuff". Parker does not deny this distinction but contends that its epistemological significance is overestimated. What is significant is not so much that the two systems (the experimental and the target systems) are made of "the same stuff", it is rather that there exist "relevant similarities" between the two. And being made of "the same stuff", in itself, does not always guaranty more relevant similarities between the experimental system and the target system. In the case of a traditional experiment, scientists must also justify making inferences from the experimental system to the target system.

4 Measurements

Another well-discussed kind of similarities between experiments and simulations concern the status of their outputs, and that leads us to our second issue – can the output of a computer simulation count as a *measurement*? Philosophers of science provide various and sometimes conflicting answers, depending on how they characterize measurements.

Morrison (2009) offers an interesting take on the issue by focussing on the role of models in a measurement process. Models do not only play a role when it comes to the interpretation of the outputs of an experiment; the measurement process itself involves a combination of various kinds of models (models of the measuring apparatus, correction models, models of data, etc.). This close connection between models and experiment is commonly acknowledged by philosophers of science. But Morrison (2009) goes one step further by adding that models themselves can function as "measuring instruments". To ground her claim, Morrison gives the example of the use of the physical pendulum to measure the local gravitational acceleration at the surface of the Earth. In that case (as in many other experiments), a precise measuring of the parameter under study (here the local gravitational acceleration) requires the application of many corrections (taking the air resistance into account for instance). So that many other, sometimes complex models are used, in addition to the simple model of a pendulum, to represent the measuring apparatus in an appropriate way. In other words, says Morrison, "the ability of the physical pendulum to function as a measuring instrument is completely dependent on the presence of these models." And she concludes: "That is the sense in which models themselves also play the role of measuring instruments" (2009, 35). To reinforce her point, Morrison gives another, more intricate example of measurement, where the role of models is even more central. In particle physics, when measuring a microscopic property such as the spin of an electron or the polarization of a photon, Morrison stresses that the microscopic object being measured is not directly observed. On the one hand, there is a model of the microscopic properties of the target system (the electron or the photon). On the other hand, an extremely complex instrument is used together with a theoretical model describing its behaviour by a few degrees of freedom interacting with those of the target system. And this is the comparison between these models that constitutes the measurement (Morrison 2009, 43). So for Morrison, in this kind of experimental settings, models function as a "primary source of knowledge", hence their status of "measuring devices". This extension of the notion of measuring goes hand in hand with the downplaying of the epistemological significance of material interaction with some real-world system: "Experimental measurement is a highly complex affair where appeals to materiality as a method of validation are outstripped by an intricate network of models and inferences" (Morrison 2009, 53).

Dropping the traditional emphasis on material interaction as characterizing experiment allows Morrison to contend that a computer simulation can also be considered as a measurement device. For once you have acknowledged the central role played by models in experimental measurement, striking similarities, claims Morrison, appear between the practice of computer simulation and experimental practice. In a computer simulation, you also start with a mathematical model of the real-world target system you want to investigate. Various mathematical operations of discretization and approximation of the differential equations involved in the mathematical model then give you a discrete simulation model that can be translated into a computer programme. Here too, as in a physical experiment, tests must be performed, in that case on the computer programme, to manage uncertainties and errors and make sure that the programme behaves correctly. In that respect, the programme functions like an apparatus in a traditional physical experiment (Morrison 2009, 53). And those various similarities, according to Morrison, put computer simulations epistemologically on a par with traditional experiment: their outputs can also count as measurements.

Not everybody agrees though. Giere (2009) for instance readily acknowledges the central role played by models in traditional experiment but rejects Morrison's extension of the notion of measuring on the ground that the various correcting models involved in the measurement process remain abstract objects that do not interact causally with the physical quantity under study (in Morrison's pendulum example, the Earth's gravitational field). And that suffices to disqualify them as measuring device. The disagreement thus seems to boil down to divergent views on the necessity of having some causal interaction with a physical quantity to qualify as a measurement. Consider a computer simulation of the solar system (another example discussed by Morrison and Giere). Do the outputs of this simulation (say, the positions of Saturn over the past 5,000 years) count as measurements? The laws of motion of the planets being very well established, the values provided by the simulation are no doubt more precise and accurate than the actual measurements performed by astronomers. Are they nevertheless only calculations and not measurements? It seems that legitimate answers and arguments can be given on both sides, depending on what you think is central to the notion of measurement. If you give priority to the epistemic function of a measurement, that is, providing reliable information of the values of some parameters of a physical system, then

Morrison's proposed extension seems appealing (provided that the reliability of the information can be correctly assessed in the case of a simulation – I will come back on this important issue later). But if you give priority to the ontological criteria stipulating that a measurement must involve some kind of causal interaction with a real-world physical quantity, then Morrison's proposition will seem too far-fetched. In any case, the very existence of this debate indicates a first way in which computer technology shapes scientific practice: the growing use of computer simulations directly bears on what it means to perform a measurement in science.

Another well-debated issue concerns the similarities – or lack thereof – between the way the output of an experiment is deemed reliable and the way the output of a computer simulation is. And that will lead us to the general issue of assessing the reliability of a simulation.

5 Internal Validity (Verification)

As briefly mentioned earlier, management of uncertainties and errors and calibration are essential components of a simulation enterprise. Simulationists must control for instance errors that might result from the various transformations the initial equations must go through to become computationally tractable (e.g. discretization), or errors resulting from the fact that a computer can store numbers only to a fixed number of digits, etc. And, as Winsberg (2003, 120) puts it: "developing an appreciation for what sorts of errors are likely to emerge under what circumstances is as much an important part of the craft of the simulationist as it is of the experimenter". Drawing on Alan Franklin' work (1986) on the epistemology of experiment, Winsberg adds that several of the techniques actually used by experimenters to manage errors and uncertainties apply directly, or have direct equivalents, in the process of sanctioning the outputs of a simulation. For instance, simulationists apply their numerical techniques on equations whose analytical solutions are known to check that they produce the expected results, just as experimenters use a new piece of experimental apparatus on well-known real-world systems to make sure that the apparatus behaves as expected. Also, simulationists may build different algorithms independently and check that they produce similar results when applied on the same mathematical model, just as experimenters use different instrumental techniques on a same target (say, optical microscopes and electronic microscopes) to establish the reliability of the techniques.

These various strategies aim at increasing our confidence in what is often called the *internal* validity or the *internal* reliability of a computer simulation. The point is to ensure that the solutions to the equations provided by the computer are close "enough" (given the limits put by computing power) to the solutions of the original equations. When refering to these checking procedures, scientists usually talk of *verification*. But verification is only (the first) half of the story when one wants to assess the reliability of a computer simulation. The other half, usually called *validation*, has to do with the relationship between the simulation and the real-world target system whose behavior it purports to investigate. And assessing this *external* validity will depend on what kind of knowledge about the target system you expect from the simulation.

6 External Validity (Validation)

A distinction similar to the traditional distinction between an instrumentalist view of the aims of a scientific theory and a realist one can be made about simulations. By instrumental aims, I mean here the production of outputs relative to the past (retrodictions) or future (predictions) *observable* behaviour of a real-world system. Retrodictions are very common in "historical" natural sciences such as astrophysics, cosmology, geology, or climatology, where computer simulations are build to produce data about past states of the simulated system (the spatial distribution of galaxies one billion years after the Big Bang, the position of the continents two billion years ago, the variation of the average temperature at the surface of the Earth during the Pliocene period, etc.). A very familiar example of predictions made by computer simulation is of course weather forecast. Realist aims are – it is no surprise – epistemically more ambitious. The point is not only to get empirically adequate outputs; it is also to get them for the right reasons. In other words, the point is not only to save the phenomena (past or future) at hand, it is also to provide reliable knowledge on the underlying constituents and mechanisms at work in the system under study. And this realist *explanatory* purpose faces, as we shall see, specific challenges. These challenges are more or less dire depending on how the simulations relate to well-established theoretical knowledge, and especially their degree of 'compositionality', that is the degree to which they are built from various theories and bits of empirical knowledge.

6.1 Duhemian Problem

At one end of the compositionality spectrum, you find computer simulations built from one piece of well-established theoretical knowledge (for instance computer simulations of airflows around wings built from the Navier-Stoke equations). In most cases, the models that are directly "read-off" a theory need to be transformed to be computationally tractable. And, depending on the available computer resources in terms of speed and memory, that involves idealizations, simplifications and, often, the deliberate introduction of false assumptions. In the end, as Winsberg (2003, 108) puts it, "the model that is used to run the simulation is an offspring of the theory, but it is a mongrel offspring". Consequently, the computer simulation does not simply inherit the epistemic credentials of its underlying theory and establishing its reliability requires comparison with experimental results. The problem is that when the simulated data do not fit with the experimental data, it is not always clear what part of the transformation process should be blamed. Are the numerical techniques the source of the problem or the various modelling assumptions made to get a computationally tractable model? As noticed by Frigg and Reiss (2009, 602–603), simulationists face here a variant of the classical Duhemian problem: something is wrong in their package of the model and the calculation techniques, but they might not know where to put the blame. This difficulty, specific to computational models as opposed to analytically solvable models, is often rephrased in terms of the inseparability of verification and validation: the sanctioning of a computer simulation involves both checking that the solutions obtained are close "enough" to the solutions of the original equations (verification) and that the computationally tractable model obtained after idealization and simplification remains an adequate (in the relevant, epistemic purpose-relative aspects) representation of the target system (validation), but these two operations cannot always, in practice, be separated.

6.2 The Perils of Accidental Empirical Adequacy[3]

At the other end of the compositionality spectrum lie highly *composite* computer simulations. By contrast with the kind of simulations just discussed, yielded by a single piece of theoretical knowledge, composite computer simulations are built by putting together various submodels of particular components and physical processes, often based on various theories and bits of empirical knowledge. Composite computer simulations are typically built to mimic the behavior of real-world "complex" phenomena such as the formation of galaxies, the propagation of forest fires or, of course, the evolution of the Earth climate. Typically, this kind of simulations combines instrumental and realist aims. Their purpose is minimally to mimic the observable behaviour of the system, but often, it is also to learn about the various underlying physical components and processes that give rise to this observable behaviour.

Composite computer simulations face specific difficulties when it comes to assess their reliability, in addition to the verification issues common to all kinds of computational models. The main problem, I will contend, is that the empirical adequacy of a composite simulation is a poor guide to its representational adequacy, that is, to the accuracy of its representations of the components and processes actually at work in the target system. Let me explain why by considering how they are elaborated throughout time. Building a simulation of a real-world system such as a galaxy or the Earth climate involves putting together submodels of particular components and physical processes that constitute the system. This is usually done progressively, starting from a minimal number of components and processes, and then adding features so that more and more aspects of the system are taken into account. When simulating our Galaxy for instance, astrophysicists

---

[3]This section (and the following) directly draws on Ruphy (2011).

started by putting together submodels of a stellar disc and of a stellar halo, then added submodels of a central bulge and of spiral arms in order to make the simulations more realistic. The problem is that the more the simulation is made realistic, the more it incorporates various submodels and the more it will run into a holist limitation of its testability. The reason is straightforward: a composite simulation may integrate several inaccurate submodels, whose combined effects lead to predictions conformed to the observations at hand. In other words, it is not unlikely that simulationists get the right outcomes (i.e. in agreement with the observations at hand), but not for the right reasons (i.e. not because the simulation incorporates accurate submodels of the actual components of the target system). And when simulationists cannot test the submodels independently against data (because to make contact with data, a submodel often needs to be interlocked with other submodels), there is unfortunately no way to find out if empirical adequacy is accidental. Therefore, given this pitfall of accidental empirical conformity, the empirical success of a composite computer simulation is a poor guide to the representational accuracy of the various submodels involved.

6.3 Plasticity and Path Dependency

Looking at simulation building processes reveals other, heretofore underappreciated features of composite computer simulations that also directly bear on the issue of their validation, to wit, what I have called their path-dependency and their plasticity. Let me (briefly) illustrate these notions with the example of a simulation of the evolution of our universe.[4] As is well known, cosmology starts by assuming that the large-scale evolution of space-time can be determined by applying Einstein's field equations of gravitation everywhere. And that plus the simplifying hypothesis of spatial homogeneity, gives the family of standard models of modern cosmology the "Friedmann-Lemaître" universes. In itself, a Friedmann-Lemaître model cannot account for the formation of the cosmic structures observed today, in particular the galaxies: The "cold dark matter" model is doing this job. To get off the ground, the cold dark matter model requires initial conditions of early density fluctuations. Those are provided by the inflation model. This first stratum of interlocked submodels allows the simulation to mimic the clustering evolution of dark matter. Other stratums of submodels, linking the dark matter distribution to the distribution of the visible matter must then be added to make contact with observations.

The question that interests us now is the following: at each step of the simulation-building process, are alternative submodels with similar empirical support and

---

[4]My discussion is based on an analysis of the Millennium run, a cosmological simulation run in 2005 (Springel et al. 2005), but similar lessons could be drawn from more recent ones such as the project DEUS: full universe run (see www.deus-consortium.org). Accessed 22 June 2013.

explanatory power available? And the (short) answer is yes (see Ruphy (2011) for a more detailed answer based on what cosmologists themselves have to say). Moreover, at each step, the choice of one particular submodel among other possibilities constrains the next step. In our example, inflation, for instance, is appealing only once a Friedmann-Lemaître universe is adopted (which requires buying a philosophical principle, to wit, the Copernican principle). When starting, alternatively, from a spherically symmetric inhomogeneous model, inflation is not needed anymore to account for the anisotropies observed in the cosmic microwave background. So that the final composition of the simulation (in terms of submodels) turns out to depend on a series of choices made at various stages of the simulation building process.

A straightforward consequence of this path-dependency is the *contingency* of a composite simulation. Had the simulationists chosen different options at some stages of the simulation building process, they would have come up with a simulation made up of different submodels, that is, with a different picture of the components and mechanisms at work in the evolution of the target system. And the point is that those alternative pictures would be equally plausible in the sense that they would also be consistent both with the observations at hand and with our current theoretical knowledge. To deny this would clearly partake of an article of faith. Path-dependency puts therefore a serious limit to the possibility of *representational* validation, that is, to the possibility of establishing that the computer simulation integrates the right components and processes.

Plasticity is another (related) source of limitation. Plasticity refers to the possibility of adjusting the ingredients of a simulation so that it remains successful when new data come in. Note, though, that plasticity does not boil down to some ad hoc fine-tuning of the submodels involved in the simulation. Very often, the values of the free-parameters of the submodels are constrained independently by experiment and observation or by theoretical knowledge, so that the submodels and the simulation itself are progressively "rigidified". Nevertheless, some leeway always remains and it is precisely an essential part of the craft of the simulationist to choose which way to go to adjust the simulation when new data come in.[5] It is therefore not possible to give a general analysis of how these adjustments are achieved (they depend on the particular details specific to each simulation building process). Analysis of actual cases suggests, however, that the way a composite simulation is further developed in response to new data usually does not alter previously chosen key ingredients of the simulation.[6] Hence the *stability* of the simulation. In other words, there is some kind of inertial effect: one just keeps going along the same modelling path (i.e. with the same basic ingredients incorporated at early stages), rather than starting from scratch along a different modelling path. This inertial effect should come as

---

[5] See for instance Epstein and Forber (2013) for an interesting analysis of the perils of using macrodata to set parameters in a microfoundational simulation.

[6] This is the case for instance for the astrophysical and cosmological simulations discussed in Ruphy (2011) and for the Earth climate simulations analyzed in Lenhard and Winsberg (2010).

no surprise, given the pragmatic constraints on this kind of simulation enterprise. When a simulation is built over many years, incorporating knowledge from various domains of expertise, newcomers do not usually have the time nor the competences to fully investigate alternative modelling paths.

The overall lesson is thus the following: because of its plasticity and its path dependency, the stability and empirical success of a composite computer simulation when new data come in cannot be taken as a reliable sign it has achieved its realist goal of representational adequacy, i.e. that it has provided accurate knowledge on the underlying components and processes of the target system (as opposed to the more modest instrumental aim of empirical adequacy).

Let us take stoke here of the main conclusions of the previous epistemological discussions.

Computer simulations may fail or succeed in various ways, depending on their nature and on our epistemic expectations. We have seen that sanctioning a computer simulation involves minimally verification issues. Those issues might be deemed more of a mathematical nature than of an epistemological nature.[7] In any case, they are clearly directly dependent on the evolution of computing power and technology. Then come the validation issues, that is, sanctioning the relationship between the computer simulation and the real-world system whose behaviour it purports to mimic. A first level of validation is empirical: do the outputs of the simulation fit with the data at hand? In most cases, however, simulationists are not merely seeking empirical adequacy, they also aim at representational adequacy. The two are of course interdependent (at least if you are not a die-hard instrumentalist): empirical adequacy is taken as a reliable sign of representational adequacy, and representational adequacy justifies in its turn trusting the outputs of a simulation when the simulation is used to produce data on aspects of a real-world system for which observations or measurements are impossible (say, the radial variation of temperature at the centre of the Earth). When assessing empirical adequacy, simulationists may face a variant of the Duhemian problem: they might not be able to find out where to put the blame (on the calculation side or on the representational side) when there is a discrepancy between real data and simulated data. Sanctioning the representational adequacy of an empirically successful simulation may be even thornier, especially for composite computer simulations. For we have seen that, because of the path-dependency and the plasticity that characterize this kind of simulations, the more composite a simulation gets to be more realistic (i.e. to take into account more aspects and features of the system), the more you loose control of its representational validation. In other words, there seems to be a trade-off between the realistic ambition of a simulation and the reliability of the knowledge it actually delivers about the real components and processes at work in the target system.

For all that, taking the measure of these validation issues should not lead to a dismissal of the scientific enterprise consisting of developing purportedly realistic simulations of real-world complex phenomena. Rather, it invites to reconsider the

---

[7]See the exchange on this topic between Frigg and Reiss (2009) and Humphreys (2009).

epistemic goals actually achieved by these simulations. My main claim is that (empirically successful) composite computer simulations deliver *plausible* realistic stories or pictures of a given phenomenon, rather than reliable insights on what is *actually* the case.

7 Modal Knowledge

Scientists (at least epistemologically inclined ones) often warn (and rightly so) against, as the well-known cosmologist George Ellis puts it, "confusing computer simulations of reality with reality itself, when they can in fact represent only a highly simplified and stylized version of what *actually* is" (Ellis 2006, 35, My italics). My point is, to paraphrase Ellis, that computer simulations can in fact represent only a highly simplified and stylized version of what *possibly* is. That models and simulations tell white lies has been widely emphasized in the philosophical literature: phenomena must be simplified and idealized to be mathematically modelled, and for heuristic purpose, models can also knowingly depart from established knowledge. But the problem with composite computer simulations is that they may also tell non-deliberate lies that do not translate into empirical failure.

The confusion with reality feeds on the very realistic images and videos that are often produced from simulated data, thanks to very sophisticated visualization techniques. These images and videos "look" as they had been obtained from observational or experimental data. Striking examples are abundant in fields such as cosmology and astrophysics, where the outputs of the simulations are transformed into movies showing the evolution of the structures of the universe over billions of years or the collision of galaxies. In certain respects, the ontological status of this kind of computer simulations is akin to the status of richly realistic novels, which are described by Godfrey-Smith (2009, 107) as talks about "sets of fully-specific possibilities that are compatible with a given description".

The stories or pictures delivered by computer simulations are *plausible* in the sense that they are compatible both with the data at hand and with the current state of theoretical knowledge. And they are *realistic* in two senses: first because their ambition is to include as many features and aspects of the system as possible, second because of the transformation of their outputs into images that "look" like images built from observational or experimental data. I contend that computer simulations do produce useful knowledge about our world to the extent that they allow us to learn about what could be or could have been the case in *our* world, if not knowledge about what is or was actually the case in our world. Note that this *modal* nature of the knowledge produced by simulations raises resistance not only among philosophers committed to the idea that scientific knowledge is about *actual* courses of events or states of affairs, but also among scientists, as expressed for instance by the well-known evolutionary biologist John Maynard Smith "[. . .] I have a general feeling of unease when contemplating complex systems dynamics. Its devotees are practicing fact-free science. A fact for them is, at best, the output of a computer simulation: it is rarely a fact about the world" (1995, 30).[8]

This increasing modal nature of the knowledge delivered by science via the use of computer simulations is not the only noticeable general transformation prompted by the development of computing power. Also on a quite general note, it is worth investigating to what extent ever increasing computing power, by stimulating the building of increasingly detailed simulations of real-world phenomena, shape the very *aims* of science.

8 Shaping the Aims of Science: Tentative Concluding Remarks

Explanation is often considered as a central epistemic aim: science is supposed to provide us with explanatory accounts of natural (and social) phenomena. But do the growing trend of building detailed simulations mean more and better explanations? There is no straightforward answer to that question, if only because philosophers disagree on what may count as a good scientific explanation and what it means for us to understand a phenomenon. Some indicative remarks may nevertheless be made here. Reporting a personal communication with a colleague, the geologist Chris Paola wrote recently in *Nature*: "... the danger in creating fully detailed models of complex systems is ending up with two things you don't understand – the system you started with, and your model of it" (2011, 38). This quip nicely sums up two kinds of loss that may come with the increasing "richness" of computer simulations.

A much-discussed factor contributing to the loss of understanding of a simulation is "epistemic opacity". Epistemic opacity refers to the idea that the computations involved in many simulations are so fast and so complex that no human or group of humans can grasp and follow them (Humphreys 2009, 619). Epistemic opacity also manifests itself at another level, at least in composite computer simulations. We have seen that these simulations are often built over several years, incorporating knowledge and contributions from different fields and different people. When using a simulation to produce new data or to test new hypotheses, the practitioner is unlikely to fully grasp not only the calculation processes but also the various submodels integrated in the simulation, which are then treated as black boxes.

As regards the loss of understanding of the target system, at least two reasons may be put forward to account for it. Recall first one of the conclusions of the previous epistemological discussion about validation: the more detailed (realistic) a simulation is, the more you loose control of its representational validation. So if the explanatory virtue of a simulation is taken as based on its ability to

---

[8] I borrow this quotation from Grim et al. (2013), which, in another framework, also discusses the modal character of the knowledge produced by simulation.

deliver reliable knowledge about the real components and mechanisms at work in the system, then indeed, fully detailed computer simulations do not score very high. But even though the reliability of the representation of the mechanisms at work provided by the simulation could be established, there would be another reason to favour very simplified simulations over simulations that include more detail. This is the belief that attention to what is truly essential should prevail on the integration of more details, when the simulation is built for explanatory purpose (rather than instrumental predictive purposes). Simulations of intricate processes of sedimentary geology, as analysed in Paola (2011), is a case in point. Paola (2011, 38) notes that "simplified representations of the complex small-scale mechanics of flow and/or sediment motion capture the self-organization processes that create apparently complex patterns." She explains that for many purposes, long-profile evolution can be represented by relatively simple diffusion models, and important aspects of large-scale downstream variability in depositional systems, including grain size and channel architecture, can be understood in terms of first order sediment mass balance. Beyond the technicalities, the general lesson is that "simplification is essential if the goal is insight. Models with fewer moving parts are easier to grasp, more clearly connect cause and effect, and are harder to fiddle to match observations" (Paola 2011, 38). Thus there seems to be a trade-off between explanatory purpose and integration of more details to make a simulation more realistic. If fewer and fewer scientists resist the temptation to build these ever more detailed simulations, feed on the technological evolution of computing power, explanation might become a less central goal of science. Predictive (or retrodictive) power may become more and more valued, since increasingly complex computer simulations will allow to make increasingly detailed and precise predictions, on ever finer scales, on more and more various aspects of a phenomenon.

Another general impact calling for philosophical attention is of a methodological nature: very powerful computer means make bottom-up approaches more and more feasible in the study of a phenomenon. In these approaches, the general idea is to simulate the behaviour of a system by simulating the behaviour of its parts. Examples of these microfoundational simulations can be found in many disciplines. In biology, simulations of the folding of proteins are built from simulations of amino-acid interactions; in ecology, simulations of the dynamics of eco-system are based on the simulations of preys-predators interactions, etc. This shaping of general scientific methodology by technology sparks sometimes vivid discussions within scientific communities: bottom-up approaches are charged with reductionist biases by proponents of more theoretical, top-down approaches. This is especially the case for instance in the field of brain studies. There has been a lot of hostility between bottom-up strategies starting from the simulation of detailed mechanisms at molecular level and studies of emergent cognitive capabilities typical of cognitive neurosciences. But discussions may end in the future in a more oecumenical spirit, given the increasing ambitious epistemic aim of brain simulations. Or at least it is what is suggested by our opening example, the European top-priority HBP project (Human Brain Project), whose aim of building multiscale simulations of neuromechanisms explicitly needs general theoretical principles to move between different levels of description.[9] The HBP project is also representative of the increasingly *interactive* character of the relationship between epistemic aims and computer technology. Simulationists do not only "tune" their epistemic ambitions to the computing power available: technological evolutions are anticipated and integrated into the epistemic project itself. The HBP project for instance includes different stages of multiscale simulations, depending on computing power progress. Big simulation projects such as the HBP or cosmological simulations also generate their own technological specific needs, such as supercomputers that can support dynamic reconfigurations of memory and communications when changing scale of simulation, or new technological solutions to be able to perform computing, visualization and analysis simultaneously on a single machine (given the amount of data generated by the simulations, it will become too costly to move the generated data to other machines to perform visualization and analysis).[10] That epistemic progress is directly linked to technological progress is of course nothing new, but the ever increasing role of computer simulations in science makes the two consubstantial to an unprecedented degree.

References

Barberousse, A., Franceschelli, S., & Imbert, C. (2009). Computer simulations as experiments. *Synthese, 169*, 557–574.
Dowling, D. (1999). Experimenting on theories. *Science in Context, 12*(2), 261–273.
Ellis, G. (2006). Issues in the philosophy of cosmology. http://arxiv.org/abs/astro-ph/0602280. (Reprinted in the *Handbook in Philosophy of Physics*, pp. 1183–1286, by J. Butterfield & J. Earman, Ed., 2007, Amsterdam: Elsevier)
Epstein, B., & Forber, P. (2013). The perils of tweaking: How to use macrodata to set parameters in complex simulation models. *Synthese, 190*, 203–218.
Franklin, A. (1986). *The neglect of experiment*. Cambridge: Cambridge University Press.
Frigg, R., & Reiss, J. (2009). The philosophy of simulations: Hot new issues or same old stew? *Synthese, 169*, 593–613.
Giere, R. N. (2009). Is computer simulation changing the face of experimentation? *Philosophical Studies, 143*, 59–62.
Godfrey-Smith, P. (2009). Models and fictions in science. *Philosophical Studies, 143*, 101–126.
Grim, P., Rosenberger, R., Rosenfeld, A., Anderson, B., & Eason, R. E. (2013). How simulations fail. *Synthese, 190*, 2367–2390.
Guala, F. (2005). *The methodology of experimental economics*. Cambridge: Cambridge University Press.
Humphreys, P. (2004). *Extending ourselves*. New York: Oxford University Press.
Humphreys, P. (2009). The philosophical novelty of computer simulation methods. *Synthese, 169*, 615–626.

---

[9] As attested by the fact that the HBP project will dedicate some funds to the creation of a European Institute for Theoretical Neuroscience.

[10] I draw here on documents provided by the Human Brain Project at www.humanbrainproject.eu. Accessed 25 June 2013.

Lenhard, J., & Winsberg, E. (2010). Holism, entrenchment, and the future of climate model pluralism. *Studies in History and Philosophy of Modern Physics, 41*, 253–262.

Morgan, M. (2005). Experiments versus models: New phenomena, inference, and surprise. *Journal of Economic Methodology, 12*(2), 317–329.

Morrison, M. (2009). Models, measurement and computer simulation: The changing face of experimentation. *Philosophical Studies, 143*, 33–57.

Norton, S., & Suppe, F. (2001). Why atmospheric modeling is good science. In C. Miller & P. N. Edward (Eds.), *Changing the atmosphere: Expert knowledge and environmental governance* (pp. 67–105). Cambridge: MIT Press.

Paola, C. (2011). Simplicity versus complexity. *Nature, 469*, 38.

Parker, W. (2009). Does matter really matter? Computer simulations, experiments, and materiality. *Synthese, 169*, 483–496.

Ruphy, S. (2011). Limits to modeling: Balancing ambition and outcome in astrophysics and cosmology. *Simulation and Gaming, 42*, 177–194.

Schweber, S., & Wächter, M. (2000). Complex systems, modelling and simulation. *Studies in History and Philosophy of Science Part B: History and Philosophy of Modern Physics, 31*, 583–609.

Smith, J. M. (1995). Life at the edge of chaos? *New York Review of Books, 42*(4), 28–30.

Springel, V., et al. (2005). Simulations of the formation, evolution and clustering of galaxies and quasars. *Nature, 435*, 629–636.

Winsberg, E. (2013). Simulated experiments: Methodology from a virtual world. *Philosophy of Science, 70*, 105–125.

Chapter 8
Adopting a Technological Stance Toward the Living World. Promises, Pitfalls and Perils

Russell Powell

Abstract In this essay, I explore the theoretical, methodological and ethical dimensions of adopting a technological stance toward the natural living world. In Part 1, I discuss the importance of adaptive match as a central explanandum of biology, offer a tentative definition of "biological design," and argue that inferences of intentional design in nature often flow from rational deliberative faculties, rather than solely or even primarily from cognitive biases toward teleological explanation or culturally inculcated religious beliefs. In Part 2, I examine the virtues of technological thinking in biology that flow from important structural similarities between organisms and artifacts, which permit the testing of evolutionary hypotheses and reveal the physical constraints on evolved design. This analysis is balanced in Part 3 by an investigation of the pitfalls associated with technologic thinking in biology and in popular science education, where I discuss a range of problems that arise from thinking of organisms as machines and describing their features in artifactual terms. Finally, in Part 4, I consider ethical misgivings about embracing the technological stance, such as the worry that an 'instrumentalist' attitude toward nature could lead to the mistreatment of beings with moral status, or that the design of organisms for human purposes expresses disrespect for living things or a pernicious desire for mastery over nature.

I would like to thank the National Humanities Center, the American Council of Learned Societies, and the Templeton Foundation Grant #43160 for support of this research.

R. Powell (✉)
Department of Philosophy and Center for Philosophy and History of Science, Boston University, Boston, MA, USA

National Humanities Center, Durham, NC, USA
e-mail: powell@bu.edu

1 Introduction

Despite the astounding success of modern evolutionary theory in explaining the origins and persistence of functional complexity in nature, design thinking remains ubiquitous in biological science, education and science journalism. Organisms are frequently described and investigated as if they were rationally designed artifacts. Ecological engineering analyses play critical roles in assessing the functionality of structures, testing adaptive hypotheses, and understanding the biomechanical constraints that underwrite convergent evolution in distant lineages. Without recourse to technological thinking, it is difficult to make sense of organismic features, strategic evolutionary interactions, and the adaptive match between organism and environment. Creative biological sciences, such as synthetic biology, aim to use engineering principles to design living artifacts that are exquisitely tailored to human purposes, causing further conceptual enmeshing of evolved organism and engineered artifact. Why does technological thinking continue to feature so prominently in biological science and communication despite the profound etiological and synchronic dissimilarities between organisms and artifacts? Is this an unfortunate legacy of pre-Darwinian theories of nature that should be eliminated from or relegated to the margins of scientific discourse? Or is it a defensible, perhaps even indispensable, component of biological research and education?

In this chapter, I explore the theoretical, methodological and ethical dimensions of adopting what I will call a *technological stance* toward the natural living world. My aim is to sketch a comprehensive, accessible, and overarching view of the philosophical landscape, rather than to investigate any particular dimension in great detail. In **Part 1**, I discuss the importance of adaptive match as a central biological explanandum, offer a tentative definition of "biological design," and argue that inferences of intentional design in nature often flow from rational deliberative faculties, rather than solely or even primarily from cognitive biases toward teleological explanation or culturally inculcated religious beliefs. In **Part 2**, I examine the virtues of technological thinking in biology, which, I argue, stem from important structural similarities between organisms and artifacts that are inferentially rich, permit the testing of evolutionary hypotheses, and reveal the physical constraints on evolved design. This sanguine analysis is balanced in **Part 3** by an investigation of the pitfalls associated with technological thinking in biology and in popular science education. Here I discuss a range of problems that arise from thinking of organisms as machines and describing their features in artifactual terms. Finally, in **Part 4**, I consider ethical misgivings about embracing the technological stance, such as the worry that an 'instrumentalist' attitude toward nature could lead to the mistreatment of beings with moral status, or that the design of organisms for human purposes expresses disrespect for living things or a pernicious desire for mastery over nature.

2 Part 1: The Conceptual and Theoretical Foundations of Biological Design

2.1 That Mystery of Mysteries

Immanuel Kant (1790) famously proclaimed that there would never be a Newton for the blade of grass.[1] Kant was skeptical not only of our ability to explain the spontaneous origin of living things from inanimate matter, but also of the possibility of explaining the origins of 'natural ends' without recourse to an intelligent designer (1790/2007, 228). Many authors have been quick to tout Charles Darwin as precisely such a 'Newton,' insofar as Darwin offered the first and only naturalistic solution to what the philosopher Sir John Herschel called "that mystery of mysteries"—the origin and extinction of species. In fact, it is not clear that Darwin solved Herschel's mystery, since the role of natural selection in speciation and extinction remain contested. Darwin did, however, solve another, perhaps even more profound, biological mystery, which we might call "that Mystery of Mysteries" (in caps): Namely, the exquisite match between the traits of organisms and the ecological design problems that they need to solve.

Darwin's theory of natural selection offered an elegant mechanistic explanation of the natural adapting of means to ends that Kant claimed was in principle unsolvable. Accounting for the non-accidental pairing between the traits of organisms and their particular lifeways is one of the singular crowning achievements of evolutionary biology (Ayala 2007), unifying a staggeringly diverse set of observations under a single schema (Kitcher 1985; Brandon 1990). But Darwin was not simply another 'Newton.' Whereas Newtonian physics has been superseded by relativity theory, Darwin's postulated combination of blind variation and natural selection remains to this day the only viable explanation for the origins and maintenance of adaptive match, functional complexity, and teleological behavior in nature.

Some adaptive matches can plausibly be explained as simple coincidence. Once, while hunting for fossil shark teeth in the hilly phosphate mines of Aurora, North Carolina, I came across a small population of pale gray-colored grasshoppers that were a spot-on match for the chalky excavated substrate, providing a near-perfect camouflage from birds. Thinking that I had discovered a case of 'industrial albinism' (adaptive lightening to human-altered environments), I snatched up one of the grasshoppers and brought it back to an entomologist at Duke University, where I was completing my doctoral work at the time. As it turned out, to my surprise, the Aurora mine grasshoppers were actually invaders from a remote coastal population. As luck would have it, they blended near-perfectly into the excavated substrate.

---

[1] The quote from Kant's *Critique of Judgment* (1790/2007, 228) reads as follows: "we may confidently assert that it is absurd for human beings even to entertain any... hope that maybe another Newton may some day arise, to make intelligible to us even the genesis of but a blade of grass from natural laws that no design has ordered. Such insight we must absolutely deny to mankind."

In contrast to such 'single-dimension' organism-environment pairings, traits that have been molded along multiple dimensions to solve a complex adaptive problem cannot plausibly be explained away as a fluke. Following Allen and Bekoff (1995) and Lewens (2004), I will use the phrase "biological design" to describe any product of cumulative selection that has been 'shaped' or 'molded' along multiple, coordinated dimensions to produce a complex function or adaptive match—an outcome that, in paradigmatic cases, is astronomically unlikely to have arisen through pure chance or stochastic processes alone.

For example, some species of butterfly in their larval (caterpillar) state mimic snakes in order to discourage predation by insectivorous birds. One such butterfly, the spicebush swallowtail (*Papilio troilus*), boasts a suite of morphological and behavioral modifications that result in an uncanny resemblance to the common green snake. This includes a thorax in the shape of a snake's head featuring two large snake-like eyes, a red retractable 'forked tongue' (which is actually a pheromone-emitting organ), and a rearing/striking behavior that mimics the aggressive posturing of snakes.[2]

All biological designs are functional in the sense that they proliferated in a population due to their fitness-enhancing effects (see Sect. 2.2). However, not all functional traits rise to the level of biological designs. Some traits originate from a single mutation and are swept to fixation by selection without being shaped along multiple, developmentally independent, dimensions. For instance, if the extant white polar bear coat originated in a single mutation that was driven to fixation by selection, it would not constitute "biological design" on the definition given above despite having evolved to solve a straightforward ecological *design problem*—namely, avoiding visual detection against a pale substrate. Hence, neither selection nor adaptive match is sufficient for biological design.[3] Furthermore, as in the realm of artifacts, not all configurations that are properly referred to as "designs" will be functional in the sense of constituting a straightforward adaptive match. Some sexually selected traits, such as ornamental features like the peacock's elaborate tail or the lizard's dewlap, constitute *adaptive mismatch by design*—these are thought to be selectively shaped predation 'handicaps' that send hard-to-fake signals of vitality to prospective mates.

It is best to think of biological design as a continuum: the greater the number of developmentally independent parameters of a trait that are shaped through cumulative selection, the more that trait will tend to resemble paradigmatic cases

---

[2]Creationists have been keen to point to putative 'irreducibly complex' traits in which a change to any trait parameter would allegedly vitiate the functionality of the trait. But most biological designs are not delicate in this respect. The spicebush swallowtail morphology, for instance, would still have bird-averting properties even if it lacked the retractable forked 'tongue'.

[3]Note, however, that selective 'shaping' should not be construed solely in topological terms—it only requires selection for a trait with multiple, independently modifiable parameters that can be represented in a phase space. If the evolution of polar bear coats involved selection along multiple, developmentally independent parameters, then it would constitute an instance of biological design.

of biological design, such as the vertebrate eye. Interestingly, the same holds true for artifacts: the more that features of an object work together in coordinated fashion to produce a specialized utility (or aesthetic outcome), the clearer that object is one of design. A simple flake struck from a rock core is far less obviously a case of design than is a samurai sword.

2.2 Design Without a Designer

In ordinary language, something's being designed implies that it has a designer. The concept of 'design' is thus infused with intentionality, planning and purpose. In contrast, what I have been calling "biological design" is the product of blind variation and natural selection.[4] Yet, Darwin's enduring mechanistic solution to the problem of adaptive match did little to banish design concepts and other teleofunctional language from biology. Unlike the physical and chemical sciences, biology remains entrenched in what Dennett (1995) has called a "design stance" toward the natural living world. Consequently, some have read Darwinian theory not so much as exorcizing teleology from biology, but rather as vindicating it by providing a theoretically sound foundation for the explanatory role of functions in biology.

The long-standing puzzle surrounding biological function was this: How could the function of a trait explain that trait's existence, when the laws of physics require—*contra* Aristotle—that causes precede their effects? The Darwinian solution to this puzzle was to say that function talk is simply shorthand for a causal-etiological claim about the history of selection for some effect. So, for example, to say that the function of the vertebrate heart is to circulate blood is to say that the vertebrate heart exists in its present form and at its present frequency because its tendency to circulate blood had fitness-enhancing effects on ancestors that possessed hearts (Neander 1991). One great virtue of an account of function that is indexed to a history of selection for effects is that it renders biological functions explanatory without violating physical law and without adverting to purposes or intentions.

Importantly, our ability to detect or intuit biological design does not depend on our ability to detect or intuit histories of selection. We identify biological design qua explanandum irrespective of the explanation that it is ultimately afforded. Moreover, we are quite capable of recognizing a biological structure as one of design without understanding its specific function. For instance, we may know that the bony plates on the back of the dinosaur *Stegosaurus* were selected to serve some function, but

---

[4]Dawkins (1997) suggests that we refer to natural objects of apparent design as "designoids," in order to distinguish these from genuine objects of design like artifacts. Similarly, Ruse (2004, 265) suggests that at the very least, we should refer to biological design as "seemingly organized complexity." In this paper, I use the phrase "biological design" unless otherwise qualified to mean the type of organized complexity produced by natural selection acting on blind variation.

not know which particular function that was. Was it to regulate body temperature, serve as armor, produce a colorful threat display, attract mates, or some combination of these effects?

The independence of design attributions from design explanations raises an interesting question: should our concept of "biological design" be indexed to selective etiology, much like the prevailing concept of function, or should it remain mechanism- and hence explanation-neutral? The definition of biological design given in the preceding section adverts to a history of selective shaping. But one might argue that it is a mistake to incorporate the *explanans* (cumulative selection) into the explanandum (biological design), lest the explanans fails to explain. If this is correct, then it seems that we should prefer a concept of biological design that does not entail any particular explanans, and which remains distinct from our concept of biological function and our ability to impute specific functions.

2.3 The Cognitive Foundations of Biological Design Attributions

How is it that we come to recognize natural design without knowing anything about the processes that gave rise to it? Do the same cognitive faculties implicated in the recognition of objects qua artifacts also play a role in the identification of biological design? While there are no definitive answers to these questions, a growing body of psychological research indicates that design thinking in relation to the living world is closely connected to the specialized faculties that subserve cognition in the domain of human artifacts.

Aristotle held that all things have a 'final cause'—a purpose or reason for existing. On this view, rain exists to nourish plants, plants exist to nourish grazing animals, and grazing animals exist to nourish humans. Pre-school children might aptly be described as natural Aristotelians in that they are inclined to attribute purposes not only to artifacts and living things, but also to inanimate natural objects such as clouds, mountains and streams. Deborah Kelemen (1999) refers to this tendency as 'promiscuous teleology'. Kelemen argues that promiscuous teleology is a byproduct of other cognitive adaptations, such as an innate bias toward agency detection. Her idea is that humans have an innate, adaptive capacity to make inferences about the goals, intentions and purposes of agents and artifacts, and that promiscuous teleology results from this capacity being extended to epistemically unwarranted domains, such as objects whose behavior can be explained by reference to purely physical (non-mental) causes. Most educated people abandon the teleological stance toward inanimate objects by the time they reach adulthood. In contrast, perceptions of the living world often remain teleological throughout life, reflecting (on Kelemen's view) overactive agency detection faculties that are recalcitrant to scientific education.

It is not clear, however, that overactive agency detection is the whole story when it comes to attributions of biological design. There is evidence that preschool children possess a specialized capacity for teleofunctional reasoning about organisms that is distinct from, and not a simple extension of, their ability to reason about agents and artifacts (Atran 1998). For example, by second grade, children tend to judge that features of a plant exist for the good of the plant, whereas they prefer physical explanations for the colors of gemstones (Keil 1994; Kelemen 2003). Furthermore, young children distinguish the 'internal' teleology of organisms from the 'external' teleology of artifacts. They judge that the features of an artifact are good for the maker or user of the artifact, rather than for the good of the artifact itself (Hatano and Inagaki 1994). For example, a thorn on the stem of a rose is judged good for the rose itself, whereas a barb on a string of barbed wire is judged good for its external human user.

It is ultimately unclear whether reasoning about organisms implicates the cognitive faculties that are implicated in reasoning about agents and artifacts. What is clear is that teleofunctional thinking plays a specialized, adaptive role in how humans reason about living things. For instance, young children use functional information—rather than overall similarity cues—to make inferences about the lifeways and behaviors of animals. In one fascinating study, Kelemen et al. (2003) presented preschool-age children with images of two insects—one a beetle (with small mandibles) and one an ant (with large mandibles)—and told them that the beetle hides from dangerous animals whereas the ant fights off dangerous animals. The researchers then presented the children subjects with a third image of a beetle with large mandibles, and asked them whether they think that it hides from or fights off dangerous animals. Despite the overall similarity between the two beetle images, the children overwhelmingly responded that the novel animal fights off dangerous animals—and where possible they offered functional justifications for that inference. From a selectionist standpoint this is not surprising, since functional information is often more predictive of organismic behavior than is overall similarity. In the language of contemporary biology, we can often infer more information about trophic position and behavioral ecology from a functional analysis than we can from a phylogenetic analysis (i.e., from genealogical relatedness). We can conclude from the formidable jaws of *Tyrannosaurus rex* that the animal behaved in ecologically important ways more like a tiger than it did like a brontosaur, despite its greater overall similarity to the latter.

Design thinking in relation to artifacts was adaptive, presumably, because it allowed us to predict how these devices interacted with other objects in the world. Similarly, by focusing on functional traits like teeth, horns, armor, camouflage, sensory apparatuses, and other 'inferentially rich' structures, early humans gained instant access to a wealth of ecologically relevant information about the probable behaviors of organisms. The clear adaptive value of teleofunctional reasoning about organisms makes incidental byproduct explanations look less compelling.

2.4 The Logical Foundations of Biological Design Attributions

To what extent can we say that inferences of intentional biological design emanate from deliberative mental processes, as opposed to intuitive faculties or lower-level cognitive biases? Note that to claim that design attributions are 'reason-based' in this sense is different from claiming that such attributions are ideally rational or epistemically justified, all-things-considered. Reasoning processes may give rise to fallacious inferences despite being rational in the non-ideal sense. Nevertheless, one way of approaching this question is to consider whether there are *any* epistemic contexts in which the intentional design inference is ideally rational. As it turns out, there is a long-standing philosophical debate over whether intentional design would be a rational explanatory inference in relation to the natural living world *if* there were no viable scientific alternative on offer.

William Paley (1802) imagined walking along a deserted shrubland and coming across a watch. Unlike a rock, whose nature and position can be explained as the outcome of chance processes, a watch contains numerous specialized parts exquisitely arranged so as to perform a particular function—an incredibly unlikely configuration that cries out for intentional explanation. Paley reasoned that like watches and other complex human artifacts, organisms are organized so as to produce precise specialized functions that could not plausibly result from a pure chance process. Paley was correct insofar as he held that chance or stochasticity is not a plausible explanation of ubiquitous functional complexity. If we compare Paley's design argument to the 'pure chance' hypothesis, it looks very attractive. But if we compare it instead to the modern Darwinian alternative, it loses much of its force. Darwin's great insight was that heritable traits vary in ways that affect organism-environment pairings, resulting in fitness differences between competing variants. Natural selection is precisely a *non-random* sampling process, even if the variation on which selection works is generated randomly with respect to its adaptive value. The explanatory virtues of evolutionary theory over intelligent design 'theory' have been extensively documented, and I will not rehearse them here (for discussions, see Dawkins 2009; Shanks 2004).

Skepticism of intelligent design preceded Darwinian theory. David Hume, for example, contended that the argument for intelligent design fails on logical grounds even though no credible mechanistic alternative had, at the time of his writing, been discovered. According to Hume, the argument for design is an argument from analogy between organisms and artifacts, which proceeds as follows: we observe a similarity in *structure* between organisms and artifacts—in particular, both exhibit a "curious adapting of means to ends"—and then we infer from this structural similarity to a similarity in *origins*. Since we know that artifacts originate in the plans and intentional actions of rational beings, we likewise infer that organisms originate in the plans and actions of rational being(s) whose product "resembles... though it much exceeds, the productions of human contrivance" (Hume [1779] 1947, 143). Hume argued that this argument from analogy fails because the organism is a fundamentally different kind of thing than even the most complex human artifact.

Hume's reply misses the mark because organisms and artifacts could differ substantially in many respects (e.g. growth) even while both exhibit an exquisite functional arrangement that is astronomically unlikely to have arisen by chance alone. In essence, Hume's objection to the design argument falls flat because, as Elliott Sober (2004) has shown, the most charitable interpretation of the design argument is not an analogical one, but rather a probabilistic inference to the best explanation. Sober glosses this inference in terms of a comparative likelihoods approach, which takes the following form: the design hypothesis is a better explanation than the chance hypothesis of some biological observation O (e.g. the vertebrate eye) if the probability of O given that the design hypothesis is correct is greater than the probability of O given that the chance hypothesis is correct.[5]

Sober argues that a fundamental problem with the design argument, even in the absence of a scientific alternative, is that it relies on certain implicit auxiliary assumptions about the causal powers or folk psychological properties of the alleged designer—assumptions that we are not permitted to feed into the likelihood assessment absent some independent line of evidential support. When intelligent design proponents point to the elegant construction of the vertebrate eye and intelligent design skeptics point to avoidable imperfections in the same (e.g. the blind spot), both parties are relying on assumptions to which they are not entitled—namely the desires, goodness, etc. of the designer—and then running an inference to the best explanation based in part on these unwarranted auxiliary assumptions. Thus, Sober concludes, we cannot say that observed biological design is more likely given the special creation hypothesis than it is given the chance hypothesis. If Sober is right, then Paley's argument for design foundered on logical grounds long before Darwin came along with a successful alternative.

Sober's analysis raises a serious problem not only for intelligent design theory, but also for its most prominent critics, many of whom have appealed to the ubiquity of sub-optimal design in nature as among the most powerful evidence against intelligent design. Sober (2007, 4) refers to this as the "no designer worth his salt" argument. Darwin's observations of sub-optimal design in nature motivated his skepticism of special creation, causing him to exclaim: "What a book a Devil's Chaplain might write on the clumsy, wasteful, blundering low [and] horridly cruel works of nature!" (quoted in Dawkins 2003). Stephen Jay Gould has defended this view, arguing that

> Ideal design is a lousy argument for evolution, for it mimics the postulated action of an omnipotent creator. Odd arrangements and funny solutions are the proof of evolution—paths that a sensible God would never tread but that a natural process, constrained by history follows perforce (1992, 21).

---

[5]Sober's 'likelihood' reconstruction holds that "Observation O favors intelligent design over chance if and only if Pr (O/ID) > Pr (O/Chance)" (2004, 122). He adopts the likelihood approach over Bayesian methods because the latter require that we assign prior probabilities to intelligent design and chance, respectively, which could skew the analysis.

Sober's point, though, is that for all we know sub-optimal design could be the result of the quirky aesthetic preferences of the designer, or perhaps even a supernatural expression of humor. Given the inscrutability of the creator, all possible desires could be built into such auxiliary hypotheses, none of which can be independently confirmed, and which can be gerrymandered to achieve the desired result. This gerrymandering has the effect of rendering intelligent design theory untestable in principle.

Imagine, however, that we discovered a patch of DNA in all known organisms in which "Made by God" was spelled out in Hebrew letters using DNA nucleotides. Sober's argument implies that the 'the stamp of Yahweh' observation is no more likely due to intelligent design than it is due to chance, since introducing any auxiliary hypotheses about what a designer would be likely to do (e.g., the Creator would autograph or trademark His creations) would be contrived. But this does not seem right. The stamp looks like a clear-cut confirmation of the design hypothesis, even if intelligent design theorists do not take the absence of such a stamp as disconfirmation of their theory, and even if we cannot independently justify claims about the desires of the Creator. Imagine that in addition to possessing the stamp of Yahweh, biological design was far more elegant than that which is actually observed, and that there was no fossil record of transitional forms to speak of. Boudry and Leuridan (2011) argue, quite plausibly, that in such a fantastical case the design argument could in theory unify a range of observations and make testable predictions that render it superior to theoretical alternatives.

Perhaps the ideally rational position, were no viable scientific theory on hand, would be to conclude that there is in all probability a mechanistic law-like explanation for the generation of biological design that continues to elude us. As Lewens (2004, 163) notes, "it is better to remain content, as Hume did, with the mystery of adaptation... than [to introduce] an intelligent designer who designs through mysterious means." Although Kant remained agnostic to the causes of biological teleology, he asserted that the inference of intelligent design is rationally compelling and "squares with the maxim of our reflective judgment" (1790/2007, 228). Indeed, there is empirical support for the notion that inferences of intelligent design emanate from biological design attributions, rather than the reverse. The most frequent justification educated people give for intelligent design-leaning beliefs is that features of the living world appear as if they were purposefully designed. Michael Shermer and Frank Sulloway surveyed members of the Skeptics Society—a highly educated and scientifically informed sample—and found that around 30 % of self-identified skeptics who believe in God (a purposeful, higher intelligence that created the universe) do so because the living world appears as if it were well-designed (Shermer 2002).

There are, no doubt, many people who accept special creation as an empirically insensitive matter of faith. My point, however, is that biological systems present, so powerfully, as objects of design that even scientifically informed and logically minded people are liable to infer agentic forces behind their production. The inference of intentional design often flows from thoughtful deliberation and cannot be explained away as an example of promiscuous teleology or religiously motivated dogma. Next, I will show that the same structural similarities between organisms and artifacts that motivate attributions of intelligent design also make technological thinking useful in biological science.

3 Part 2: Technological Thinking in Biology: The Promise

Thinking of organisms in technological terms is central to inferential reasoning in ecology and evolution, as well as to the formulation and testing of selection hypotheses. Technological thinking plays an indispensable role in understanding biomechanical systems and the physical constraints on organismic design. And it is responsible for major 'weight-bearing' in the logical structure of evolutionary theory.

3.1 Organisms as Artifacts: The Case of Darwin's Moth

Technological thinking in biology can lead to specific predictions in ecology and evolution which, when vindicated, represent paradigmatic confirmations of the theory of natural selection. Consider the following case of an unusual moth hypothesized by Charles Darwin. Darwin was intrigued by a Malagasy orchid that kept its nectar at the bottom of a 30 cm-long trumpet-like structure. At the time, no insect was known that could pollinate such a flower. Noting this ecological design problem, Darwin drew upon a theory of coevolution between orchids and their insect pollinators to predict the existence of a giant hawkmoth with an improbably long proboscis (Micheneau et al. 2009).

The proboscis is a retractable, pipette-like structure through which insects lap up nutrient-rich fluids, such as nectar. Darwin hypothesized an evolutionary arms race between the length of the orchid spur and the length of the proboscis: where the proboscis is longer than the orchid spur, the moth is able to extract nectar without pressing its head firmly against the orchid and thus without pollinating it; conversely, where the proboscis is shorter than the orchid spur, the moth will be less capable of extracting nectar and thus less likely to attempt nectar extraction and thus less likely to act as a pollinator. The result of this competitive interaction is a lock-and-key fit between the proboscis and the orchid spur.

From the hypothesis of strategic interaction and careful observations of pollination activities, Darwin was able to make precise predictions about the length of the hypothesized proboscis and the organism to which it would be attached. Darwin's hypothesized moth was discovered many years after his death (it was named *Xanthopan morganii praedicta*, after Darwin's prediction), and was not conclusively implicated as the pollinator in question for another century. Thinking of the proboscis as a 'tool' that is optimized by natural selection for extracting a desired resource, and of the orchid spur as responding in kind to 'technological advances' in the proboscis, was undoubtedly a significant cognitive factor in the formulation of Darwin's hypothesis. The very notion of an evolutionary 'arms race' or 'strategic interaction' conjures an image of rational technological move and counter-move, albeit without intentionality and played out over evolutionary time. The example of Darwin's moth could be multiplied many times over, suggesting that technological thinking is not merely conceptual shorthand, but rather plays a substantive cognitive role in organizing adaptive hypotheses (see Sect. 3.3).

Technological thinking in biology has borne not only epistemic but also technological fruit. The field of 'biomimetics' seeks to emulate biomechanical solutions to common ecological design problems, and to put these solutions to work for human ends. For instance, researchers at Clemson University are using the moth proboscis as an engineering prototype for designing 'bioinspired' devices that can be used for probing, transporting and controlling liquids droplets of varying viscosity (Vatansever et al. 2012). The first telescope to significantly reduce chromatic aberration, introduced by English inventor Chester Moore Hall in the eighteenth century, was modeled on the structural solution deployed in the vertebrate eye. Presently, researchers are looking to the visuo-structural adaptations of mantis shrimp to improve synthetic polarizing optics, which are currently significantly outperformed by 'natural' biophysical solutions (Roberts et al. 2009). The field of robotics has long attempted to emulate insect mechanical structures and control architectures, albeit with limited degrees of success due more to matters of physical scale than to the limits of technological thinking (Ritzmann et al. 2004). In short, artifacts serve as models for understanding organisms and their evolution, and organisms serve as models for designing and improving artifacts.

3.2 Two Worlds of Design

The case of Darwin's moth, and countless similar examples, show that in important respects, the worlds of organisms and artifacts are both worlds of design—and moreover, that it is fruitful to investigate them as such. There are profound differences, of course, between artifacts and organisms due to fundamental differences in the processes that produce them (Sect. 4.1). But analogies are not to be regarded as true or false—rather, they are more or less useful for inferring, understanding, explaining or predicting the properties of objects. Analogical reasoning between organisms and artifacts is useful because it taps into important non-accidental similarities between these kinds. For example, the cambered wing foil of birds and airplanes (discussed below) is not a coincidental similarity—it is a robust structural result of physical constraints on flight acting in conjunction with a function-optimizing process.

Given the non-accidental similarities between natural and artifactual design, it is understandable that many reasonable people would infer the existence of a biological designer in the absence of a credible alternative explanation. Somewhat more surprising is that organism-artifact comparisons did not cease after Darwinian theory garnered widespread acceptance. In some areas of contemporary biology, thinking of organisms as engineered artifacts is cognitively indispensable. The field of biomechanics, for example, sets out to investigate the structure of 'living technology'. According to Steven Vogel, Duke University biologist and pioneer of the field, "life forms a technology in every proper sense, with a diversity of designs, materials, engines, and mechanical contrivances of every degree of complexity" (1998, 16). Organisms and artifices are subject to the same physical laws, pressures, temperatures, fluid mediums, gravitational forces, and so on.

If the 'design problems' facing certain evolutionary lineages are highly similar to those facing human engineers, and if the solution to these design problems are highly constrained by physical laws, then we might expect similar solutions to emerge in natural and artifactual design despite fundamental differences in their underlying modes of production. Convergent evolution between distant lineages, such as the independent origination of camera-type eyes in vertebrates and cephalopod mollusks, indicates that there may be a limited set of evolutionary solutions to common ecological design problems (Powell 2012; McGhee 2011; Conway Morris 2003). In some cases, the physical laws impose such severe constraints on viable functional design that processes as different as mechanistic natural selection and foresighted human engineering will tend to converge on similar solutions. This appears to be the case for the problem of resolving images from electromagnetic radiation (discussed above), as well as for alternative modes of forming images from waveform energy, such as echolocation. Scientists began experimenting with active sonar well before sophisticated echolocation systems were discovered in bats and cetaceans, which were subsequently used as a model for 'bioinspired' devices.

Shared physical and environmental constraints have underwritten significant convergence between organismic and artifactual design, as recounted in Vogel (1998, 17):

> Both bicycle frames and bamboo stems take advantage of the way a tube gives better resistance to bending than a solid rod. A spider extends its legs by increasing the pressure of the fluid inside in much the same way that a mechanical cherry picker extends to prune trees or deice planes. Both [living and artifactual] technologies construct things using curved shells (skulls, eggs, domed roofs), columns (tree trunks, long bones, posts), and stones embedded in matrices (worm tubes, concrete). Both use corrugated structures... to get stiffness without excessive mass—whether the shell of the scallop... or the stiffening structures of doors, packing boxes, and aircraft floors, or fan-folded paper and occasional roofs. Both catch swimming or flying prey with filters through which fluid flows-whether spiders or whales, gill-netting fishers or mistnetting birders.

Like their human-engineered counterparts, animal wings, themselves convergent in the history of life, tend to have 'cambered' airfoils (curved wing tops with flattened bottoms) for maximal lift and minimal drag. Physical constraints on locomotion through viscous fluids have resulted in the repeated evolution of the 'fusiform' shape (a spindle form that is tapered at the ends) in fish, Mesozoic marine reptiles, marine mammals, cephalopod mollusks, and human-engineered craft both submarine and aerial.

Organism-artifact similarities go beyond overarching morphology and descend to the particular 'nuts and bolts' of biological construction. Consider Vogel's (1998, 186–7) description of the bacterial flagellum:

The base of the flagellum forms a driveshaft that passes through the cell membrane, connecting it to a rotary engine. And the membrane works like a proper set of bearings. The engine bears a curious similarity in both appearance and operation to our electric motors. It's even reversible. The whole thing—engine and corkscrew—either singly or in groups, pushes or pulls a bacterium around much the same way a propeller pushes a ship or pulls an airplane.

Such descriptions are rife with helpful technological and, in particular, mechanical metaphor. In other cases, the analogy between organisms and artifacts is more tenuous. If we define "engine" in broad functional terms as "a structure that inputs non-mechanical energy into mechanical systems," then muscles, motile cilia, and even sub-cellular organelles responsible for respiration and photosynthesis (such as mitochondria and chloroplasts) would count as "engines." Why, in any case, do we refer to mitochondria as "microscopic engines," instead of referring to engines as "macroscopic mitochondria?" The answer seems straightforward and straightforwardly cognitive: we use familiar concepts to understand, relate to, and convey information about the properties of unfamiliar objects. It is this cognitive phenomenon that gives technological thinking in biology its purchase.

3.3 Biological Value of the Technological Stance

We have seen how technological thinking is of great utility in formulating adaptive hypotheses, picking apart the casual structure of biomechanical systems, and identifying the physical constraints on the evolution of form. A number of authors (e.g. Dennett 1995; Ruse 2004) have defended what Tim Lewens (2002/2004) calls the "artifact model of the organism." This is a policy suggestion for biological inquiry recommending that biologists investigate organisms as though they were artifacts, despite the significant dissimilarities between these ontological domains. The artifact model is motivated by a fundamentally pragmatic claim about the best way for biologists to unpack the causal structure of the living world. It is not motivated by a claim about how organisms are *in fact* constructed. It may turn out that organisms are composed of many useless and cumbersomely interconnected parts—but the idea behind the artifact model is that we approach organisms *as if* they have reasonably well-designed structures that are crafted for specific, isolatable functional roles.

'Methodological adaptationism' (*sensu* Godfrey-Smith 2001) is the thesis that biological inquiry is most fruitfully conducted when organismic traits are approached as if they are functional. The 'technological stance' associated with the artifact model implies more than simply investigating traits as if they were selected for particular purposes. It cognitively capitalizes on approaching organismic features as if they were technical artifacts. For instance, biologists modeling the functional moth proboscis (discussed above) describe this structure as a "drinking straw," the fluid uptake properties of which depend on the action of a "sucking pump" in the moth's head (Monaenkova 2011). The drinking straw model of the moth proboscis, like so many similar models in biology, relies heavily on an analogy to human artifacts for its cognitive purchase.

The technological stance structures our thinking about the nature of adaptive design. According to Michael Ruse, "for the natural theologian, the heart is literally designed by God— metaphorically, we compare it to a pump made by humans. For the Darwinian, the heart is made through natural selection, but we continue, metaphorically, to understand it as a pump made by humans" (Ruse 2004, 265). Vogel's defense of the technological stance is even stronger: "We've only rarely recognized any mechanical device in an organism with which we weren't already familiar from engineering" (1998, 311). If this is correct, then the technological stance will be vital to investigations in anatomy, physiology and cell biology, as well as to our understandings of evolution and ecology. It is only by using technological frames of reference that allow us to deploy concepts from more familiar domains of human experience that we are able to make sense of the staggeringly complex and unfamiliar causal structure of the organism.

Technological thinking also plays an important theoretical role in the structure of evolutionary theory. The philosopher Herbert Spencer famously characterized the process of natural selection as 'survival of the fittest' (without objection from Darwin), from which a logical problem ensued. If we define relative fitness in terms of reproductive success—wherein the fittest organisms are those that survive and reproduce—then we have rendered the principle of natural selection tautologous and hence non-explanatory. There is a voluminous and sophisticated literature in the philosophy of science addressing the 'problem of fitness' from numerous angles. One of the more promising strategies for avoiding the tautology problem equates fitness with probabilistically expected (rather than actual) reproductive success. The fittest are not those organisms that survive and reproduce, but those that have relatively higher values of expected reproductive success. This allows us to distinguish differential reproduction ('sampling') that is due to fitness differences ('selection') from differential reproduction that is due to other factors ('drift'). The technological stance enters into the picture when we attempt to assign relative fitness values and appeal to these differences in order to explain differential reproduction.

Biologists often determine relative fitness values through the perspective of an 'ecological engineer.' Dennett (1995) has proposed that we define relative fitness in terms of the ability to solve design problems set by the environment. The design problem cannot, of course, be reproductive success simpliciter, as this would be to slip back into the Spencerean tautology. Biologists must instead specify the *ecological design problems* that a particular organism needs to solve—e.g., what it eats, how it forages, what predators it must avoid, how it reproduces, and so on— and from this information determine the relative fitnesses of competing variants in a population.

Fitness values are a function of the relation between organismic features and environmental properties, and no trait is fit in all environments—for example, sometimes being bigger or smarter is advantageous, and sometimes these traits are disadvantageous. For this reason, fitness must be assessed on a case-by-case basis, and the principle of natural selection will only admit of testable predictions once the ecological details are filled in and the engineering analysis takes shape (Brandon 1990, 20). Thus, it is mainly by recourse to technological thinking that the causal connection between heritable variation and differential reproduction is hypothesized and, through careful observation, established. In fact, it is by way of a pre-theoretical engineering analysis that people come to recognize the ubiquity of adaptive match in the first place.

4 Part 3: Technological Thinking in Biology: The Pitfalls

4.1 *Organisms as Machines*

Despite the seemingly indispensable value of technological thinking in biology, adopting a technological stance toward the living world can also lead to serious problems. The notion that we should conceive of organisms as machines has roots in Descartes' philosophy of nature. In *Principles of Philosophy* (1644), he wrote "The only difference I can see between machines and natural objects is that the workings of machines are mostly carried out by apparatus large enough to be readily perceptible by the senses (as is required to make their manufacture humanly possible), whereas natural processes almost always depend on parts so small that they utterly elude our senses" (quoted in Vogel 1998, 40). According to Descartes, animals, as well as human bodies, are machines whose microscopic inner working can in theory be understood in mechanical terms. On the Cartesian view, organisms are not *like* machines—they *are* a type of machine.

Insofar as machine thinking implies only mechanistic as opposed to vitalistic causes, there is little basis for objection. However, when machine thinking engenders comparisons between organisms and what we might call "quintessential machines," such as automobiles and computers, one can begin to see the grounds for concern. Modern molecular biology is replete with quintessential machine imagery, with biological analogues of factories, assembly stations, engines, motors, pistons, pumps, blueprints, software programs, and the like. The worry is that such analogies will be extended beyond their domain of utility or, even worse (though less plausibly), transformed into an identity relation (Recker 2010).

All technical artifacts are machines in the sense that they are used to modify force. However, quintessential machines are more complex than simple machines (such as levers and wedges), and as a result of this complexity they exhibit interesting similarities to organisms. Both quintessential machines and organisms are composed of specialized parts and part types; both exhibit complexly organized functions; and both can behave teleologically. Yet there are profound differences between organisms and quintessential machines that run the risk of being obscured by machine thinking in biology.

Perhaps the most significant difference between organisms and quintessential machines lies in their internal organizational dynamics (Nicholson 2013). Organisms are self-organizing, self-reproducing, and self-maintaining systems. No such machines currently exist. Quintessential machines, such as airplanes, may have most of their parts replaced by external human engineers over time, but they are not capable of repairing and renewing their own parts as organisms do. In addition, organismic systems respond flexibly in ontogenetic time to environmental challenges they encounter—a type of adaptive plasticity that Vogel (1998, 241) refers to as "demand-responsive alteration." For example, muscles grow in response to stress and emaciate in its absence; callouses form in response to friction and dissipate in its absence; and so on. In comparison to organisms, quintessential machines are developmentally inflexible and structurally brittle. Although some autonomous machines exhibit very simple goal-directed behavior (such as heat-seeking missiles or drones), even the teleological behavior of "simple" animals, such as insects, has no parallel among quintessential machines. In all of these respects, technological thinking will tend to detract from biological understanding, rather than illuminate it.

Furthermore, as Lewens (2004) points out, another potential methodological pitfall of thinking of organisms as artifacts is that this may cause researchers to neglect the developmental interconnections between parts of the organism. Machines tend to exhibit highly modular designs that allow for modification of their components without disrupting the overall functioning of the system in which they are embedded. In contrast, the selective shaping (Sect. 2.1) of organismic form is constrained by gene-gene interactions, as well as the one-to-many relations of the genotype-phenotype map—cumbersome developmental complexities that have no analogues in the quintessential machine world. This is not to say that developmental modularity is not an important factor in evolutionary innovation and diversification (e.g., the segmented body plan of insects probably played an important role in the great evolutionary success of this clade). But as Gould and Lewontin (1979) warned in their seminal critique of adaptationism, neglecting these developmental interactions can obscure the historical constraints on natural selection in shaping organismic form.

Technological thinking can also cause one to overlook differences in process that lead to important differences in product. Quintessential machines are rationally constructed with a pre-specified goal in mind, assembled with standardized materials gathered from far and wide, and rapidly improved upon in space and time. Organisms, by contrast, are the outcome of a mindless, incremental and excruciatingly slow process of natural selection, working only with the materials on hand, and tinkering with existing structures and developmental systems even when they are poorly cut out for the relevant ecological task (Sect. 2.4). Further, unlike rational engineers who are able to go back to the drawing board when a design turns out to be impracticable or inefficient, natural selection is often path-dependent and deeply constrained by history. As a result, nature will often fail to set a gold standard for artifactual design.

4.2 Encouraging Unwarranted Inferences of Intelligent Design

A final difference between organisms and artifacts relates to the origins of their respective teleologies. The functions of artifacts are indexed to the beliefs, desires or uses of *external* agents, whereas the functions of organismic features are determined by non-rational processes that are *internal* to the system in question, such as the role that such features play in system maintenance and/or reproduction. Insofar as organism-machine analogies gloss over this distinction, they could have worrisome implications for biology education. If not qualified, technological thinking could encourage or reinforce unwarranted inferences of intelligent design.

Machine-related language, which is widely used in educational materials and science journalism, can readily be coopted by contemporary creationists as 'evidence' that scientists secretly embrace intelligent design (Pigliucci and Boudry 2011; Nicholson forthcoming). Intelligent design apologists, such as Michael Behe (2006), quite intentionally refer to subcellular processes as "molecular machines" and "assembly lines" with all the intentionality that such descriptions entail. Even if, as I have argued, technological thinking is indispensable to much of biological science, biologists should nevertheless pay heed to the current political climate and the wider social implications of how they characterize and communicate their findings. In this way, epistemological values and social values can come into conflict in the practice of science.

This raises the question: Is it possible to effectively communicate biological research to the general public without appealing to the language of quintessential machines? The prospect of banishing all vestiges of design thinking from biology education and communication is both unlikely and undesirable, given the deep structural similarities between organisms and artifacts (Sect. 3.2), and given the fact that human cognition is heavily disposed toward teleofunctional reasoning (Sect. 2.3). Nevertheless, research suggests that early elementary school children are capable of sophisticated forms of biological reasoning, including with respect to the conceptual precursors of complex evolutionary concepts like common descent, extinction and speciation (Nadelson and Sinatra 2009). Given that folk biological structures emerge early in human development and remain psychologically entrenched (Kelemen 1999), it is imperative that we provide an early and accurate conceptual foundation for biology education. Having said this, it is difficult to imagine teaching biology to children—be it anatomy, evolution, ecology or behavior—without recourse to technological thinking and broader teleofunctional concepts. Nor would it be desirable to do so, given the cognitive utility of such approaches. Rather, in both education and science journalism, the limits of technological metaphors should be expressly acknowledged, and the dissimilarities between natural and intentional design—both in process and in product—should be consistently underscored.

5 Part 4: Technological Thinking in Biology: The Perils

Thus far, we have explored the promises and pitfalls of technological thinking in biology. In this final and concluding section, we will look at some ethical implications of adopting a technological stance toward the natural living world.

5.1 Technology Made Human

It is not hyperbole to say that technology made us human. Only a handful of taxa are capable of transmitting socially learned behaviors across generations, and none but *Homo* is capable of a cumulative technological industry: the innovation, improvement and transmission of technical artifacts down the generations.

The ability to fashion tools from non-living or formerly living matter, such as stone, wood, bone and hide, was a crucial factor in the transition from bipedal chimp-brained Australopithecine ancestors to early humans. Reductions of the human gut, jaws and teeth, and corresponding enlargements of the neocortex, were possible in part due to the transmission of simple technologies that allowed for the hunting and butchering of game as well as the thermal processing of food. It was not until the upper Paleolithic that augmented capacities for cumulative culture gave rise to the high-fidelity/high bandwidth transmission of cultural innovation that culminated in human behavioral modernity (Sterelny 2012). Technological capacities further expanded with the population boom and specialization of labor that followed in the wake of the Agricultural revolution (Diamond 1997), and yet again during the scientific revolution. Today technology progresses at such a breakneck pace that older generations are compelled to acquire cultural innovations predominantly from their descendants, reversing the intergenerational flow of cultural information that characterized hunter-gather human populations for over a million years.

5.2 Ethical Implications of the Technological Stance

If we are, as our evolutionary history suggests, obligatory technovores, what could be morally problematic about adopting a technological stance toward the living world? By assuming a 'Technological Stance' (all caps) toward nature, I mean something more specific, and more specifically cognitive, than simply possessing robust technological capabilities. I mean an orientation toward the natural world that conceives of living things primarily in terms of their utility to humans. The Technological Stance is similar to what Heidegger called "Ge-stell," which is often translated as "enframing" (Zimmerman 1990). According to Heidegger, Ge-stell reveals living things as an undifferentiated pool of resources—a standing reserve of function ("Bestand") to be manipulated in ways that are instrumental to human ends. Natural objects present anonymously as things to shape and control; they are thereby "ontologically subordinated" to human function, and as a result their non-technical nature is concealed.

Is this a plausible description of modern scientific attitudes toward nature and, if so, is it morally objectionable? Heidegger tended to focus on physics because of its emphasis on quantification, universality and control. But his critique can be applied as well to the ongoing revolution in biotechnology (Schyfter 2012). 'Creative' biosciences, such as genetic engineering and in particular synthetic biology, present us with increasingly powerful ways of converting the natural world into 'living technology'. Humans have been producing living technology ever since the advent of selective breeding programs in the Agricultural Revolution. However, the new creative biosciences represent a significant break from these ancient efforts in their ability to bypass the random sexual recombination of genomes in order to carry out precise genetic modifications and even the de novo synthesis of entire genomes, promising exponentially greater control over organisms and their properties.

One important difference between synthetic biology and even the most sophisticated selective breeding programs is that it involves the application of rational engineering principles to organismic design. Instead of deriving new life forms by tinkering with existing living systems, synthetic biology aims to design organisms from the ground (or minimal microbial platform) up by compiling and drawing upon a standardized registry of biological parts. These 'building blocks' can be combined in numerous ways to produce organisms that are exquisitely tailored to human purposes (Endy 2005; O'Malley 2009). This engineering approach could make synthetic biology particularly susceptible to the Heideggerian critique, to the extent that, for its practitioners, "there do not exist living things; rather, there exist functions, which are transferable without limitations" (Schyfter 2012, 217).

It is clearly hyperbole to suggest that synthetic biologists do not recognize organisms as anything but standing reserves of function. Surely, people can and do view living things from multiple perspectives at the same time (more on this below). But suppose, for the sake of argument, that creative biotechnologies do encourage individuals to view organisms as ontologically equivalent to the inanimate material that we use to fashion ordinary artifacts. This would indeed raise a host of ethical concerns. First, and most troubling, it could entail that the interests of beings with moral standing—such as persons and sentient beings more broadly—are not taken into account as reasons for or against human action. The interests of factory-farmed animals are already disregarded in this way, insofar as agricultural animals are treated entirely as economic units of production. Perhaps the Technological Stance we are inclined to take toward animal domesticates partially stems from their pseudo-artifactual status. One worry, then, is that technological approaches toward the living world could exacerbate attitudes that encourage individuals to treat beings in ways that are inconsistent with their moral status. Secondly, adopting a Technological Stance toward non-sentient organisms could be morally problematic, not because it entails the direct neglect of any morally protectable interests, but because the technological manipulation of non-sentient beings could have harmful consequences for, or disvalue to, beings that do have moral standing. An example would be engaging in technological manipulations of the living world that lead to global ecosystem or climate disruptions.

But neither the mistreatment of beings with moral status, nor the dis-valuable manipulation of non-sentient living things, is *inherent* to the creative biological enterprise (Douglas et al. 2013). There is no plausible psychological basis to support the Heideggerian view that people cannot engage in technological manipulations of the living world while at the same time recognizing the intrinsic value and non-technical nature of the subjects of manipulation. The fact that in developed countries scientific experiments on human and non-human animals is justified, restricted, and prohibited on moral grounds, shows that we are capable of treating beings with moral standing at the same time both as moral subjects that are valuable in themselves, and as means to other valuable ends.

If we take the Technological Stance to be a pervasive, unqualified and all-consuming approach to the living world, such that nearly all of our interactions with living things are approached from the standpoint of utility to the neglect of genuine subjects of moral worth, then it is deeply morally problematic. But viewed in this way, it is clearly a straw man—an attitude that virtually no people, and certainly very few biologists, exhibit. Biologists are often among the most ardent conservationists, many having been drawn to biology out of a deep admiration for natural design—not a desire to remake or master nature. Many scientists are in the business of modifying naturally existing organisms with the hopes of finding treatments for disease, or ameliorating anthropogenic effects on climate change (*cf.* Buchanan 2011). The notion that biologists or biotechnological engineers have grown decreasingly mindful of the moral value of morally valuable beings is a sweeping and empirically implausible claim.

5.3 Does Nature Deserve Respect?

One might argue that engineering organisms to suit human ends is inherently disrespectful of nature, regardless of whether or not it disregards the interests of uncontroversial moral subjects (for such a view, see Boldt and Müller 2008). But just what sort of 'respect' is due to non-rational, non-sentient entities is unclear.

There is currently no widely accepted theory of moral status that gives significant moral weight to the purely biological interests of non-sentient beings (such as bacteria), or to non-individual collectivities (such as communities and ecosystems). Furthermore, the ethic of respect is closely tied to Kantian moral theory and the central importance such theories assign to autonomy, practical rationality, dignity, and consent. Because respect is tightly linked to rational agency and associated concepts, it is only tenuously applied to nonrational sentient beings, and not at all to non-sentient organisms. One might operate with a radically different notion of respect than is found in contemporary moral philosophy, such as, e.g., a broader "reverence for nature." But in that case, the onus is on the proponent of such a definition to make a principled distinction between laudable or permissible biological interventions, such as treating an infection with antibiotics or selectively breeding crops for human consumption, and interventions that are allegedly inconsistent with the principle of respect for nature, such as the genetic modification or synthesis of organisms for the very same purposes. To my knowledge, no such principled distinction has been made.

In short, ontological enframing in the Heideggerian sense would be ethically problematic if it obscured the morally relevant properties of living things and thus caused us to treat beings in ways that are inconsistent with their moral status. But, as we have seen, there is little reason to think that technological attitudes toward non-sentient organisms must, or are likely to, lead to the disregard of morally relevant interests, or to the mistreatment of genuine subjects of moral worth.

6 Conclusion

In conclusion, the technological stance is of great theoretical, methodological, and cognitive value to various subfields of biology and biotechnology. If adopted overzealously or without qualification, however, it can cause us to overlook certain ontological or etiological properties of the organism or to falsely infer others. In extreme forms, a technological approach could encourage a perception that the living world is merely a standing reserve of function to be converted without limit into technology suited to human ends. Yet the veneration of nature, which compels us to tout biological design as the epitome of engineering excellence and to view human interventions in the genetic fabric of the living world as inherently disrespectful, is no antidote for the pitfalls and perils of technological thinking. For it, too, is a form of ontological enframing that conceals from view the botched and amoral character of natural design—perhaps the best evidence we have that Darwin solved, once and for all, that Mystery of Mysteries.

Acknowledgments I am grateful to Irina Mikhalevich for extremely thorough and helpful comments on an earlier draft of this essay. I am grateful to the National Humanities Center's Robert and Margaret Goheen Fellowship, the American Council of Learned Societies, and to Templeton Foundation Grant # 43160, as well as the Center for Genetic Engineering and Society at North Carolina State University, for financial support of this research.

References

Allen, C., & Bekoff, M. (1995). Biological function, adaptation, and natural design. *Philosophy of Science, 62*(4), 609–622.

Atran, S. (1998). Folk biology and the anthropology of science: Cognitive universals and cultural particulars. *Behavioral and Brain Sciences, 21*, 547–609.

Ayala, F. J. (2007). Darwin's greatest discovery: Design without designer. *Proceedings of the National Academy of Sciences, 104*(suppl.), 8567–8573.

Behe, M. J. (2006). *Darwin's black box: The biochemical challenge to evolution* (10th anniversary ed.). New York: Free Press.

Boldt, J., & Müller, O. (2008). Newtons of the leaves of grass. *Nature Biotechnology, 26*, 387–389.

Boudry, M., & Leuridan, B. (2011). Where the design argument goes wrong: Auxiliary assumptions and unification. *Philosophy of Science, 78*(4), 558–578.

Brandon, R. N. (1990). *Adaptation and environment*. Princeton: Princeton University Press.

Buchanan, A. (2011). *Beyond humanity*. Oxford: Oxford University Press.

Conway Morris, S. (2003). *Life's solution: Inevitable humans in a lonely universe*. Cambridge: Cambridge University Press.

Dawkins, R. (1997). *Climbing mount improbable*. New York: W.W. Norton & Co.

Dawkins, R. (2003). *A Devil's Chaplain: Reflections on hope, lies, science, and love*. Boston: Houghton Mifflin Harcourt.

Dawkins, R. (2009). *The greatest show on earth: The evidence for evolution*. London: Bantam.

Dennett, D. C. (1995). *Darwin's dangerous idea*. New York: Simon & Schuster.

Diamond, J. (1997). *Guns, germs, and steel: The fates of human societies*. New York: W.W. Norton.

Douglas, T., Powell, R., & Savulescu, J. (2013). Is the creation of artificial life morally significant? *Studies in History and Philosophy of Biological and Biomedical Sciences, 44*(4), 688–696.

Endy, D. (2005). Foundations for engineering biology. *Nature, 438*, 449–453.

Godfrey-Smith, P. (2001). Three kinds of adaptationism. In S. H. Orzack & E. Sober (Eds.), *Adaptationism and optimality* (pp. 335–357). Cambridge: Cambridge University Press.

Gould, S. J. (1992). *The panda's thumb: More reflection in natural history*. New York: W.W. Norton & Co.

Gould, S. J., & Lewontin, R. C. (1979). The spandrels of St. Marcos and the Panglossian paradigm: A critique of the adaptationist programme. *Proceedings of the Royal Society of London B, 205*, 581–598.

Hatano, G., & Inagaki, K. (1994). Young children's naive theory of biology. *Cognition, 50*, 171–188.

Hume, D. (1779/1947). *Dialogues concerning natural religion*. Library of Liberal Arts.

Kant, I. (1790/2007). *Critique of judgment* (N. Walker, Ed. & J. C. Meredith, Trans.). Oxford University Press.

Keil, F. C. (1994). The birth and nurturance of concepts by domains: The origins of concepts of living things. In L. A. Hirschfeld & S. Gelman (Eds.), *Mapping the mind: Domain specificity in cognition and culture* (pp. 234–254). Cambridge: Cambridge University Press.

Kelemen, D. (1999). The scope of teleological thinking in preschool children. *Cognition, 70*, 241–272.

Kelemen, D. (2003). British and American children's preferences for teleo-functional explanations of the natural world. *Cognition, 88*, 201–221.

Kelemen, D., et al. (2003). Teleo-functional constraints on preschool children's reasoning about living things. *Developmental Science, 6*(3), 329–345.

Kitcher, P. (1985). Darwin's achievement. In N. Rescher (Ed.), *Reason and rationality in natural science* (pp. 127–189). Lanham: University Press of America.

Lewens, T. (2002). Adaptationism and engineering. *Biology and Philosophy, 17*, 1–31.

Lewens, T. (2004). *Organisms and artifacts: Design in nature and elsewhere*. Cambridge, MA: MIT Press.

McGhee, G. (2011). *Convergent evolution: Limited forms most beautiful*. Cambridge, MA: MIT Press.

Micheneau, C., Johnson, S. D., & Fay, M. F. (2009). Orchid pollination: From Darwin to the present day. *Botanical Journal of the Linnean Society, 161*, 1–19.

Monaenkova, D. (2011). Butterfly proboscis: Combining a drinking straw with a nanosponge facilitated diversification of feeding habits. *Journal of the Royal Society Interface*. doi:10.1098/rsif.2011.0392.

Nadelson, L. S., & Sinatra, G. M. (2009). Educational professionals' knowledge and acceptance of evolution. *Evolutionary Psychology, 7*, 490–516.

Neander, K. (1991). Functions as selected effects: The conceptual analyst's defense. *Philosophy of Science, 58*, 168–184.

Nicholson, D. J. (2013). Organisms ≠ machines. *Studies in History and Philosophy of Biological and Biomedical Sciences, 44*(4), 669–678.

O'Malley, M. A. (2009). Making knowledge in synthetic biology: Design meets kludge. *Biological Theory, 4*(4), 378–389.

Paley, W. (1802). *Natural theology: Or evidences of the existence and attributes of the deity.* Collected from the Appearances of Nature, Reprinted Gregg, Farnborough, 1970.

Pigliucci, M., & Boudry, M. (2011). Why machine-information metaphors are bad for science and science education. *Science & Education, 20*(5–6), 453–471.

Powell, R. (2012). Convergent evolution and the limits of natural selection. *European Journal for Philosophy of Science, 2*(3), 355–373.

Recker, D. (2010). How to confuse organisms with mousetraps: Machine metaphors and intelligent design. *Zygon, 45*(3), 647–664.

Ritzmann, R. E., Quinn, R. D., & Fischer, M. S. (2004). Convergent evolution and locomotion through complex terrain by insects. *Arthropod Structure & Development, 33*, 361–379.

Roberts, N. W., Chiou, T. H., Marshall, N. J., & Cronin, T. W. (2009). A biological quarter-wave retarder with excellent achromaticity in the visible wavelength region. *Nature Photonics, 3*(11), 641–644.

Ruse, M. (2004). *Darwin and design: Does evolution have a purpose?* Cambridge, MA: Harvard University Press.

Schyfter, P. (2012). Standing reserves of function: A Heideggerian reading of synthetic biology. *Philosophy and Technology, 25*, 199–219.

Shanks, N. (2004). *God, the devil, and Darwin: A critique of intelligent design theory.* Oxford: Oxford University Press.

Shermer, M. (2002). *Why smart people believe weird things.* Holt.

Sober, E. (2004). The design argument. In W. Mann (Ed.), *The Blackwell companion to philosophy of religion.* Malden: Wiley-Blackwell.

Sober, E. (2007). What is wrong with intelligent design? *Quarterly Review of Biology, 82*(1), 3–8.

Sterelny, K. (2012). *The evolved apprentice: How evolution made humans unique.* Cambridge, MA: MIT Press.

Vatansever, F., et al. (2012). Toward fabric-based flexible microfluidic devices: Pointed surface modification for pH sensitive liquid transport. *Applied Materials & Interfaces, 4*(9), 4541–4548.

Vogel, S. (1998). *Cats' paws and catapults: Mechanical worlds of nature and people.* New York: W. W Norton and Co.

Zimmerman, M. E. (1990). *Heidegger's confrontation with modernity.* Bloomington: Indiana University Press.

Part IV
Reflections on a Complex Relationship

Chapter 9
Goal Rationality in Science and Technology. An Epistemological Perspective

Erik J. Olsson

Abstract According to one strong intuition, what distinguish science from technology are the ultimate goals of these activities: while the goal of technology is practical usefulness, the goal of science is truth. The question raised in this paper is whether, and to what extent, this means that goal setting rationality is also different in the two domains. It is argued, preliminarily, that it is not: the theory of goal rationality in management and technology can be profitably transferred to the scientific context. This conjecture is substantiated partly by remarking on its intrinsic plausibility and partly, and above all, by appealing to its systematic advantages. As for the latter, the conjecture is applied to four closely related epistemological debates, with pragmatist ingredients, concerning truth as a goal of inquiry. It is argued that these otherwise puzzling debates can, in this way, be fruitfully reconstructed and perhaps even resolved.

1 Introduction

Authors reflecting on the distinction between science and technology often share the intuition that what demarcates the one from the other are their ultimate goals: while the goal of technology is practical usefulness, the goal of science is truth. For instance, Jarvie (1972) concludes that "[t]echnology aims to be effective rather than true" (p. 55), and Skolimowski (1972) that "[s]cience concerns itself with what is, technology with what is to be" (p. 44). The same basic intuition is clearly stated, more recently, in Houkes (2009):

> The intuition .... is that technology is, in all its aspects, aimed at practical usefulness. Thus whether technological knowledge concerns artefacts, processes or other items, whether it is produced by engineers, less socially distinguished designers, or by consumers, the prima facie reason to call such knowledge "technological" lies in its relation to human goals and actions. And just as scientific knowledge is aimed at, or more tenuously related to, the truth, so technological knowledge is shaped by its relation to practical usefulness. (p. 312)

E.J. Olsson (✉)
Theoretical Philosophy, Lund University, Lund, Sweden
e-mail: erik_j.olsson@fil.lu.se

A few paragraphs later, Houkes summarizes this intuition – the TU-intuition as he calls it – in the following terms:

> [T]he TU-intuition understands the difference between natural science and technology (or, more narrowly, the engineering sciences) in terms of a difference in goals: the former aims at finding out true theories, where the latter aims at practical usefulness. (p. 318).

Despite its many adherents, the TU-intuition has been challenged from various camps. The intuition presupposes that the kind of truth at which scientific activity is directed is not reducible to practical utility. Some instrumentalists may want to question that presupposition in favor of a "pragmatic" theory of scientific truth. From such a standpoint, the distinction between technology and science as a distinction between goals becomes difficult to uphold. Another possibility is to accept the realist concept of truth and to accept also that scientific theories are candidates for this property but to deny that scientists may justifiably accept or reject a theory because of its truth-likeness. It could be maintained, instead, that theory choice ought to be governed by the usefulness of theories for solving empirical and theoretical problems of science. If this is correct, then the proposal that technology and science differ regarding their goals becomes, once more, problematic.[1]

Yet it should be made clear that these challenges to the TU-intuition are based on rather extreme views about the notion of truth that arguably relatively few authors are willing to subscribe to, at least in unqualified terms. Moreover, the TU-intuition does not strictly speaking presuppose, in a strong sense, a realist or objectivist concept of truth. What it does assume is, again, that the kind of truth that is relevant for scientific theories is not reducible to mere practical utility. For instance, a theory which rejects any notion of scientific truth beyond empirical adequacy would still be sufficient for the purposes of underwriting the TU-intuition (so long as empirical adequacy is not itself reduced to practical utility).

Although my later discussions will shed doubt on some pragmatist proposals, I shall in the following basically take the TU-intuition for granted. My main question will rather be this: given that science and technology differ with respect to the goals that they aim at, does that also mean that they differ regarding goal *rationality*? Or is the rationality involved in setting technological goals basically the same as the rationality involved in setting scientific goals? Goal rationality has been studied extensively in management science, and some of that work has been transferred to a technological context. This has given rise to a fairly precise and well-developed framework within which technological goal rationality can be fruitfully studied. The study of goal rationality in science, by contrast, turns out to be a surprisingly underdeveloped intellectual territory.

The hypothesis to be substantiated in this paper states that the theory of goal rationality in management and technology can be profitably transferred to the scientific domain. Hence, even if the TU-intuition is correct, so that there is a

---

[1] For a review of various attempts to undermine the TU-intuition the reader is referred to the excellent exposition in Houkes (2009).

fundamental difference in the ultimate goals that science and technology aim to attain, the qualitative difference between these two phenomena should not be overestimated: they still share the same goal rationality. I will argue in favor of this conjecture partly by remarking on its intrinsic plausibility, but also partly, and above all, by appealing to its systematic advantages. Concerning the latter, I will suggest that the conjecture can be fruitfully employed in reconstructing and, at least to some extent, resolving four related epistemological debates concerning the truth as a goal of inquiry. Two of these debates are pragmatist in spirit, thus presenting potential threats to the TU-intuition along the lines suggested above. I will provide reasons for thinking that these approaches rely on an incomplete understanding of the nature of goal rationality.

2 Goal Rationality in Technology

Goal rationality has been studied extensively in management theory, where it is central in so-called MBO, an acronym standing for Management By Objectives (e.g. Mali 1972). This has led to the development of a common approach, codified in the acronym SMART, according to which goals should be Specific, Measurable, Achievable, Realistic and Time-bound. This theory has been refined and systematized by Sven Ove Hansson and his research group at the Royal Institute of Technology (KTH) in Stockholm (e.g. Edvardsson and Hansson 2005). In the following, I will refer to the framework developed by Hansson et al. as SMART+, signaling that it represents an updated, philosophically more sophisticated, version of the original SMART conditions. (This is my terminology, not theirs.) The KTH group has used the theory in its study of environmental objectives (Edvardsson 2004) and transport objectives (Rosencrantz et al. 2007). I will discuss its relevance for technology below.[2]

A goal is typically set for the purpose of achieving it. We will say that a goal is *achievement-inducing* if setting it furthers the desired end-state to which the goal refers. Thus the goal of becoming rich is achievement-inducing (for me) if my setting that goal makes it more likely that I will in fact become rich, e.g. by inspiring me to focus on accumulating wealth, which may eventually lead to my actually becoming wealthy. As a first approximation, a goal G is achievement-inducing for a subject S just in case the probability that S attains the goal G is increased by S setting herself the goal G, i.e., in semi-formal terms, just in case P(S attains the goal G | S sets herself the goal G) > P(S attains the goal G).

Edvardsson and Hansson proceed to use the notion of achievement-inducement to define the concept of goal *rationality*: in their view, a goal is rational if it performs its achievement-inducing function (sufficiently) well. This is a satisficing

---

[2]The account of SMART+ in this section draws mainly on Edvardsson and Hansson (2005). The reader is advised to consult that paper for additional references.

rather than an optimizing notion of rationality (Simon 1956). Evidently, in order to be achievement-inducing and therefore, on this proposal, rational a goal should guide as well as motivate action. One could also argue that rational goals serve to coordinate actions among several agents, but that aspect will not play any major role in the following.

There is certainly more to be said about this proposed concept of goal rationality. First, as it stands it begs the question against visionary goals such as "world peace" or, in general, goals that cannot be fully attained. An example from Swedish transport policy is the so-called "vision zero" goal stating that, in the longer run, no one should be killed or seriously injured as the effect of a traffic accident (Rosencrantz et al. 2007). A goal that cannot be attained is not achievement-inducing and hence irrational according to the proposed definition. However, there is an obvious way to avoid this untoward result by redefining achievement inducement. A goal G is achievement-inducing for a subject S, on the revised proposal, just in case the probability that S attains the goal at least partially or, alternatively, at least approaches the attainment of G, is increased by S setting herself the goal G.

Second, achievement-inducement, even in the less demanding sense, cannot be all there is to goal rationality. If it were, the rational thing to do would be to set oneself trivial goals that can be easily attained: poking one's nose, lifting one's hand, and so on. The likelihood that I manage to raise my hand if I set myself the goal to do so is very close to one. Goals which are more difficult to achieve, such as getting oneself a solid education, would be dismissed as irrational. However, the proposal does make good sense as a tie-breaking condition in a setting where there are already a number of candidate goals that have been singled out on the basis of other considerations. Faced with a set of goals that are equally attractive in other respects, it is reasonable to select one that is achievement-inducing.

With these clarificatory remarks in mind, what does it mean, more specifically, that a goal can guide and motivate action? It is useful at this point to distinguish between three types of criteria of goal-rationality: those related to what the agents *know*, what they *can do* and what they *want to do*. From the first, epistemic perspective, goals should be *precise* and *evaluable*. A goal such as "achieving a better society" fails on the first account, that of precision. That goal is not very useful for guiding action unless supplemented with more precise instructions. There are at least two different aspects of precision: directional and temporal. A goal is *directionally complete* if it specifies in what direction one should go in order to reach the goal. Take for example the goal to substantially decrease the number of unemployed in Sweden. That goal is directionally complete because it suggests in what direction progress towards the goal is to be made. If employment has decreased, then the goal has been approached or achieved, otherwise not. A goal is *temporally complete* if it specifies the timeframe within which it should be attained.

A goal is *end-state evaluable*, moreover, if it is possible to know whether it has been achieved. The goal to reduce a pollutant in the atmosphere to a certain level that is far below what can be measured would fail to satisfy the criterion of end-state evaluability. A goal is *progressively evaluable* if it can be determined how far we are from satisfying it. This property of goals is crucial in determining whether a certain course of action should be maintained, changed or given up. It has also been argued that such feedback enhances the agent's motivation so that she will make an intensified effort to act in ways that further the goal.

For an illustration, suppose my goal is to reach Geneva by the end of the day. In order for that goal to be rational, I must be able to determine whether or not this is the city I am actually in by the end of the day. However, in many situations it is not enough to be able to determine whether or not the goal state has been fully achieved. In the example, I must also be able to tell whether I am travelling in the right direction, and how far I have left to go. In particular, if a goal is distant, or difficult fully to achieve, we need to be able to judge the degree of success in approaching the goal. In other words, degrees of *partial attainment* must be distinguishable.

The second aspect of goal rationality concerns what the agent *can do*. It is reflected by the requirement that a goal should be *attainable*, or at least *approachable* (i.e. attainable at least to some degree). The goal to become a wizard (in the sense of a person with true magical powers) would not be classified as attainable or even approachable. There are at least three dimensions of approachability: *closeness*, *certainty* and *cost*. The dimension of closeness is the most obvious one. It concerns how close to the goal it is possible to come. The goal to achieve a perfectly just society is probably not fully achievable, and would therefore qualify as utopian, but it can be approached by acting in ways that increase social justice.

The third aspect of goal rationality is the volitional one. It concerns what we *want to do*. Goals, in order to be rational, should be motivating. Setting ourselves the goal should motivate us to act in a way which furthers the realization of the goal state. The motivation that a goal may give rise to in the agent can be characterized according to degree of intensity or durability. Studies indicate that goals are more action-generating when they are explicit and specific, and that such goals are more likely than do-your-best goals to intensify effort. There is also evidence suggesting that specific and challenging goals lead people to work longer at a task. We have already mentioned a connection between evaluation and motivation: when people can check how they stand in relation to a goal, their motivation to carry out the task often increases.

An insight into the nature of goal-setting emerging from SMART+ is that the criteria of rational goal-setting may conflict in the sense that the satisfaction of one criterion to a high degree may lead to a failure to satisfy substantially some other criterion. The probably most common type of such conflicts are occasioned by the fact that some of the properties that make a goal action-guiding may at the same time make it less capable to motivate action. Consider, for example, the following two goals (Edvardsson and Hansson 2005):

1. The team shall win 12 out of 20 games with a least a two goal advantage, 3 out of 20 games with at least a one goal advantage, and never lose a game with more than one goal.
2. The team shall beat all opponents hands down.

Here, the second goal, though less action-guiding than the first, is plausibly more achievement inducing, and therefore more rational, because of its greater action-motivating capacity.

In general, visionary and utopian goals are more likely to motivate action than less visionary goals, which on the other hand may be more action-guiding. The task of goal-setting therefore may very well involve a trade-off between goals that are action-motivating and goals that are action-guiding. This may lead to the formulation of one single goal reflecting this compromise. However, it is often a better idea to adopt not a single goal but a whole system of goals at different levels. As Edvardsson and Hansson point out: "One way of balancing the criteria so as to optimize goal realization is to adopt goal systems in which goals are set on different levels in order to supplement each other. In this way visionary and highly motivating goals can be operationalized through more precise and evaluable subgoals, or interim targets." (2005, p. 359)

While the SMART+ theory was developed for the purposes of studying managerial goal-setting rationality, it has been noted to be "well applicable to engineering practice" (Hughes 2009, p. 393). There is one exception to this rule, though: the motivity of a goal, the degree to which it motivates those involved to work towards it, is not much discussed in the design literature, being more naturally regarded an issue for management.[3] As for approachability, engineers pay special attention to this aspect via feasibility studies which is a standard tool for securing approachability. Precision is vital in engineering, not least in the problem definition phase. The design process starts with the identification of a need, whether it arises from a client's needs or in some other way, giving rise to the broad goal of the engineering project. However, it is necessary that this goal be clarified before any serious design work can be done. Thus, the problem definition stage is often conceived of as an essential part of the design process (e.g. Dym and Little 2004; Dieter 1983). This clarificatory process involves considering design objectives, user requirements, constraints and functions. Moreover, engineers continuously verify that the design process meets the defined goals and that the goals are reasonable, giving rise to a feedback loop that may lead to further design refinements. Hughes concludes that "like precision, evaluability is a valued feature in engineering" (ibid.).

Another aspect of goal rationality concerns the coherence of ultimate goals. We would like to have, in management as well as in technology, ultimate goals that are mutually consistent and perhaps even mutually supportive, in the sense that the fulfillment of one goal facilitates the fulfillment of another (Rosencrantz 2008). If two goals are mutually inconsistent, that means that both goals cannot be satisfied at once. Thus, a mousetrap designer would be irrational to insist that her product should not harm or restrict the freedom of the mouse in any way. But there may also be other more subtle forms of goal conflicts. Thus the satisfaction of one goal may make the satisfaction of another goal less likely. For instance, goal conflicts are common in environmental politics (Edvardsson 2004). There are two kinds of goal

---

[3]The account of goal-setting in technology that follows is based on Hughes (2009), p. 393.

conflicts within this context: internal and external. An example of an internal goal conflict is when a prohibition on using pesticides results in an increased mechanical use of land, which in turn increases the discharge of carbon dioxide. An external goal conflict arises when the expansion of the infrastructure in terms of roads and railroads threatened the preservation of sensitive biotopes. Softer forms of goal conflicts commonly arise in technology as well (Hughes 2009). To take a frequently occurring case, success in reducing cost will often enough have an adverse effect on performance, and vice versa.

Few authors would go as far as disallowing goal conflicts altogether. The more common view is to regard them as a fact of life. Thus, Edvardsson concludes that "goal conflicts seem to be unavoidable in any multigoal system, unless the goals are set very low" (Edvardsson 2004, p. 178), adding that "[t]he ideal of a perfect state of affairs in which all goals have been achieved is most likely utopian" (ibid.). It is not clear, in many such cases, that it is worthwhile to invest resources in formulating goals so that there cannot be a conflict. It may be a better strategy to allow for certain conflict potential and make the necessary trade-offs if a conflict should materialize in practice. Still, as Edvardsson points out, "it makes sense for any multigoal system to provide for some mechanism whereby goal conflicts may be solved" (ibid.). Contingency planning may be called for if conflicts are not only possible but can be expected to occur, although one should probably take into account that planning itself consumes resources. Hughes concludes, similarly, that goal conflicts should generally be allowed, noting that "[t]he [weak] consistency criterion is best reserved for those situations in which the goals are stated precisely (the cost should be no more than \$$x$ and the vehicle should accelerate at a rate of no less than y)" (Hughes 2009, pp. 392–393).

3 Goal Rationality in Science

On first sight, goal rationality in science seems attractively simple in comparison to goal rationality in technology since the goal of scientific inquiry is simply to find the truth. On closer scrutiny, however, considerable complexity emerges. For one, the goal to find the truth does not by itself suggest any very definite course of action; it does not specify in what direction one should go in order to reach the goal, except possibly that one should use a method that is reliable – one that is likely to lead to true beliefs. Still, the goal itself does not indicate what those methods are. Not only directional completeness but also the other aspects of goal rationality identified in the SMART+ model make good sense as principles governing goal rationality in the scientific domain. For another example, it would clearly be desirable in science to have a goal that is end-state evaluable in the sense that it is possible to know whether it has been achieved. Once more, the goal of truth is not an obvious candidate. Similarly, we would like scientific goals to be temporally complete, progressively evaluable, attainable, and we would be happy to have goals that exert the proper motivational force on the inquirer. Finally, there seems to be no reason to think that science is devoid of goal conflicts. For instance, the goal of truth could be satisfied by simply adopting a trivial theory, one which is logically true. To avoid this, we need the further goal of informativity. But as many epistemologists have observed (e.g. Levi 1967), if we decide to adopt both goals as ultimate ends this is likely to lead to a goal conflict since a more informative theory is often less likely to be true. A theory that is very specific regarding the causes of a particular kind of cancer may thereby be less likely to be true than a less committed theory.[4]

To add substance to these remarks concerning the prima facie structural similarity between goal rationality in science and technology I will apply the SMART+ framework to four (related) epistemological debates concerning the proper goal of (scientific) inquiry, starting with two pragmatists: Peirce and Rorty. My main aim is to indicate, without focusing excessively on interpretational details, how the SMART+ theory could inform and clarify some otherwise puzzling epistemological disputes.[5]

4 Peirce on Belief as the Goal of Inquiry

In a famous essay, Peirce argues that, contrary to the received view, the goal of inquiry is not truth, or true belief, but merely belief or "opinion" (Peirce 1955, pp. 10–11):

> [T]he sole object of inquiry is the settlement of opinion. We may fancy that this is not enough for us, and that we seek, not merely an opinion, but a true opinion. But put this fancy to the test and it proves groundless; for as soon as a firm belief is reached we are entirely satisfied, whether the belief be true or false. And it is clear that nothing out of the sphere of our knowledge can be our object, for nothing which does not affect the mind can be the motive for mental effort. The most that can be maintained is, that we seek for a belief that we shall *think* to be true. But we think each one of our beliefs to be true, and, indeed, it is mere tautology to say so.

We recall that, for Peirce, belief or opinion is, by definition, that upon which an inquirer is prepared to act. Hence, Peirce is proposing to reduce the goal of scientific inquiry to the goal of attaining that upon which we are prepared to act. Peirce's enterprise could be interpreted as presenting the TU-intuition with a direct challenge, especially if the kind of action that is referred to is practical action, a reading which is not alien to Peirce's pragmatism.

In the latter part of the quote, Peirce seems to be maintaining that the true state of things does not affect the mind and therefore cannot be the motive of mental effort. But the claim that the facts of the matter do not affect the mind is a counterintuitive

---

[4]For a related issue and some complications, see Bovens and Olsson (2002).

[5]For the purposes of simplicity and definiteness, I will in the following take "truth" in its objectivist or realist sense as referring to correspondence with an external reality, although I conjecture that much of the reasoning that follows would survive a weakening to "empirical adequacy", or the like.

one. When I look out the window, I come to believe that there is a tree just 10 m away. Normally, this belief is caused by the tree, or the fact that there is a tree, which is thus affecting my mind.[6]

On another interpretation, Peirce is thinking of objective truth as essentially "mind-independent". If so, one could be led to think that it follows trivially that objective truth cannot affect the mind, for nothing that is mind-independent can if that is what "mind-independent" means. But this is an irrelevant sense of mind-independence. In a less trivial sense, something is mind-independent and objective if it does not depend entirely on our will. Truth is mind-independent in the latter sense but not in the former. What is true – for example that there is a tree outside the window – does not depend entirely on our will but it is still something that can affect us in various ways, and typically does so through our observations.

Peirce is right, though, in stating that once we believe something, e.g. that there is a tree out there, we cannot, pending further inquiry, distinguish the state we are in from a state of *true* belief. If S believes that p, or believes truly that p, she cannot tell whether she has attained the first goal or the second. She will, from the position of the goal end state, judge that she believes that p just in case she will judge that she believes truly that p. Peirce can be understood as maintaining that this fact alone makes it more rational, or appropriate, to view the goal of inquiry in terms of fixing belief rather than in terms of fixing *true* belief. Is that correct?

Let us look at the matter from a more abstract perspective. We will say that two goals G_1 and G_2 are *end-state evaluation equivalent* for a subject S if, upon attaining one of G_1 or G_2, S cannot tell whether she attained G_1 or G_2. Peirce, in the argument under scrutiny, is relying on the following principle:

(Peirce's Principle) If (i) G_1 and G_2 are end-state evaluation equivalent for a subject S, and (ii) G_1 is logically stronger than G_2, then G_2 is more rational than G_1 for S.

Is this principle valid as a general principle of goal rationality? I will argue that it is not. Suppose that P is a pollutant that is dangerous to humans and that M is a device which indicates whether or not the amount of P in the air exceeds the limits that have been set by an international body. Moreover, there is no other device that can be used for this purpose. However, M is not fully reliable and it sometimes misfires. Let G_1 be the goal of using the device M *and* successfully determining whether the air is free of P-pollution; and let G_2 be the goal of using the device M. G_1 and G_2 are end-state evaluation equivalent for the measuring person S: upon attaining G_1 or G_2 she cannot distinguish one from the other. Moreover, G_1 is logically stronger than G_2. It would follow from Peirce's Principle that G_2 is more rational than G_1.

---

[6]It could be objected that Peirce is here using "truth" in a technical sense, signifying what is collectively accepted by all researchers once scientific inquiry has come to an end. Truth in that sense presumably does not exert any direct influence on a particular mind now. Still, this is an implausible interpretation of Peirce in the present context, as there is no concrete sign that truth should be given any special technical meaning.

But this conclusion can be questioned. It is true that G_2 is more easily attained than G_1. But G_1 is surely more inspiring than G_2; it is, to use Peirce's own expression, a stronger "motive for mental effort". It cannot, therefore, be concluded that G_2 is more rational, or achievement-inducing, than G_1. Hence, the principle presupposed by Peirce is plausibly not generally valid. This observation is sufficient to undermine Peirce's argument that the goal of belief is more rational, or appropriate, than the goal of true belief.

Indeed, the goal of true belief, or the goal of truth for short, does sound more inspirational than the goal of settling belief. Many people, not least those equipped with a scientific mind, will go to almost any length to find the truth of the matter, sometimes even in practically insignificant affairs. Disregarding the special case of religious faith, comparatively few would be willing to incur similar personal and other costs for the sole gain of settling a corresponding opinion.

Apart from the general invalidity of Peirce's Principle, there may be other differences between the goal of belief and that of true belief that are worth attending to. One such factor is a difference in precision. We recall that a goal is said to be directionally complete if it specifies in what direction one should go in order to reach the goal. We have noted that the goal of truth does not do terribly well on this score. But it might still do better than the goal of belief. For the goal of true belief suggests, albeit imperfectly, that the belief be fixed, not by any old method, but by one that is likely to establish the truth of the matter. This would suggest to the inquisitive mind such things as evidence-gathering, hypothesis-testing, the use of scientific instruments, and so on. The goal of belief does not suggest as vividly any particular course of action. It is compatible with using a wider range of methods, including methods that are not truth-oriented but focus, say, on the systematic disregard of contravening evidence.

Finally, there is a difference between the two goals on the ability dimension, concerning what we can do to approach the respective goals. This is related to the presumed difference in directional completeness. The goal of belief can be approached and evaluated along one dimension only: degree of belief. The stronger our belief is, the closer we are to achieving the goal of (full) belief. The goal of truth, by contrast, can in addition be approached, at least in principle, along the dimension of truth-likeness: the closer we are to the truth, the closer we are to achieving the goal of true belief *ceteris paribus*.

5 Rorty on Justification as the Goal of Inquiry

My second application of SMART+ concerns an argument presented by Richard Rorty in a paper from 1995, drawing partly on earlier work (e.g. Rorty 1986), to the conclusion that truth is not legitimately viewed as the goal of inquiry. This is a conclusion also drawn by Peirce, as we saw, but where Peirce thought that the

9 Goal Rationality in Science and Technology. An Epistemological Perspective

goal of truth should be replaced by the goal of belief, Rorty proposes that the proper replacement is rather *justified* belief.

The starting point of Rorty's 1995 article is the following declaration (p. 281):

> Pragmatists think that if something makes no difference to practice, it should make no difference to philosophy. This conviction makes them suspicious of the philosopher's emphasis on the difference between justification and truth. For that difference makes no difference to my decisions about what to do. If I have concrete, specific doubts about whether one of my beliefs is true, I can resolve those doubts only by asking whether it is adequately justified – by finding and assessing additional reasons pro and con. I cannot bypass justification and confine my attention to truth: assessment of truth and assessment of justification are, when the question is about what I should believe now (rather than about why I, or someone else, acted as we did) the same activity.

He adds, a few pages later on (p. 286):

> The need to justify our beliefs and desires to ourselves and our fellow agents subjects us to norms, and obedience to these norms produces a behavioral pattern that we must detect in others before confidently attributing beliefs to them. But there seems no occasion to look for obedience to an additional norm – the commandment to seek the truth. For ... obedience to that norm will produce no behavior not produced by the need to offer justification.

In arguing that the goal of scientific inquiry is not truth but being in a position to justify one's belief, Rorty is, in effect, challenging the TU-intuition, especially as he views justification as essentially unrelated to truth, which in the end is a notion he favors dropping altogether (p. 299). One of the conclusions of his essay is that, on the Dewey-inspired theory which he advocates, "the difference between the carpenter and the scientist is simply the difference between a workman who justifies his action mainly by reference to the movements of matter and one who justifies his mainly by reference to the behavior of his colleagues" (ibid.).

My ambition here is not to add to the voluminous literature on the interpretation of Rorty's pragmatism. Instead, I would like to distill one argument that I believe can be found in his essay, suitably reconstructed. Rorty, as quoted above, is contrasting two goals: the goal of attaining a true belief and the goal of attaining a justified belief. On the reading I would like to highlight, he is offering an argument that is similar to Peirce's argument for the propriety of the goal of belief, but for a slightly different conclusion. Rorty is pointing out that the goal of attaining a true belief and the goal of attaining a (sufficiently) justified belief are end-state evaluation equivalent from the point of view of the inquirer: once the inquirer has attained either of these goals, she cannot tell which one she attained. This much seems true. Yet Peirce's Principle is not directly applicable as it demands that, among the goals under consideration, one goal be logically stronger than the other. The two goals of true belief and justified belief are not at all logically related, at least not as justification is standardly conceived.[7]

Still, we note that the goal of justified belief is plausibly more directionally complete than the goal of true belief, and in the quote this is a feature that Rorty

---

[7] For more on this, see the section on Kaplan below.

highlights. On a plausible reconstruction, the general principle underlying Rorty's reasoning, then, is this:

> (Rorty's Principle) If (i) G_1 and G_2 are end-state evaluation equivalent for a subject S, and (ii) G_2 is more directionally complete than G_1, then G_2 is more rational than G_1 for S.

But this principle shares the fate of Peirce's Principle of being plausibly generally invalid. Since the problem is similar in both cases, I shall not this time give an explicit counterexample. Suffice it to note that beside directional completeness, there are several other aspects of a goal that play a part in determining its relative rationality. One such aspect is, to repeat, the motivational one. This aspect is interesting in this context because it often offsets the directional aspect. Goals that are strongly motivational are in practice rarely directionally complete, and vice versa. Thus many are motivated by goals such as achieving "world peace" or "a completely just society" and yet these goals do not *per se* suggest any particular cause of action. Conversely, goals that give detailed advice for how to act tend to be less inspirational.

As we have already noted, the goal of truth, though directionally less complete than the goal of justification, may still be more rational in virtue of its inspirational qualities. Hence, *pace* Rorty we cannot conclude, from the presumed fact that the goal of true belief and the goal of justified belief are end-state evaluation equivalent and the latter more directionally complete than the former, that the latter is also the more rational choice.

Leaving Rorty's discussion aside, a natural view to adopt concerning the relation between the two goals of true belief and justified belief, from a SMART+ perspective, is that they could very well live side by side, supplementing each other: the goal of truth providing the visionary, motivating factor and the goal of justification playing the more action-guiding part. Drawing on the upshots of Sect. 3, there are *prima facie* two ways of implementing this recommendation. One would be to adopt a system of goals wherein both goals figure, the goal of truth as a high-level goal and the goal of justification as lower-level goal, the latter operationalizing the former. The other way would be to compress the two goals into one goal, the goal, namely, to attain a justified true belief. The latter goal amounts, incidentally, to the goal of attaining *knowledge*, as that concept is traditionally conceived. Yet the claim that knowledge, in the traditional sense, is the proper goal of inquiry has been questioned by several epistemologists. I turn now to two such criticisms, due to Mark Kaplan and Crispin Sartwell, respectively.

6 Kaplan on the Irrelevance of Knowledge in Inquiry

Fifty years ago, Edmund Gettier famously argued that the traditional analysis of knowledge is mistaken (Gettier 1963). Gettier proposed two counterexamples intended to show that we may be justified in believing something which is true, without this belief qualifying as a case of knowledge in a pre-systematic sense. Gettier's paper generated an industry of attempts to solve the "Gettier problem", which is standardly interpreted as the problem of identifying additional clauses which, if they are added to the traditional account of knowledge, make Gettier's examples cases of non-knowledge. For an overview, the reader is referred to Shope (2002).

A rather different approach was taken by Mark Kaplan (1985). Rather than putting forward further clauses to supplement the traditional account knowledge, Kaplan challenged the importance of the Gettier problem as such. If it is not an important problem, then there is no pressing need to solve it. The aim of this section is to examine Kaplan's arguments more closely, and to do so specifically from the perspective of goal-setting rationality.

To fix ideas, one of Gettier's counterexamples takes the following form. Both Smith and Jones have applied for a certain job. Smith is justified in believing both that Jones will get the job and that Jones has ten coins in his pocket. Because of this, Smith also justifiably believes that the man who will get the job has ten coins in his pocket. As it turns out, unbeknownst to Smith, it is he, not Jones, who will get the job. And, as it happens, Smith, too, has ten coins in his pocket. Smith, then, has a justified true belief that the man who will get the job has ten coins in his pocket. But given the circumstances it seems that Smith lacks knowledge of this fact. If so, having a justified true belief is not sufficient for having knowledge.

As Kaplan notices, the solution to the problem may look obvious. The reason why Smith lacks knowledge of the proposition in question is surely the fact that his conclusion relies on a false premise, that, namely, Jones will get the job. Hence, the problem can be avoided by simply adding a further clause to the traditional account of knowledge ruling that, in order for a subject to know that p, the subject's reasons for p must not rely on a false premise. Unfortunately, there are similar examples which do not involve reasoning from a false premise (Goldman 1976). For our purposes, we may disregard this complication. We will follow Kaplan and assume, for the sake of the argument, that the no-false-premise solution properly handles the Gettier problem.

Now the Gettier problem is a problem for the analysis of knowledge as justified true belief only if we conceive of justification as being "fallible". In other words, we need to suppose that a subject can be justified in believing that p even if p is, in fact, false. In Gettier's example it was assumed that Smith justifiably believes the false proposition that Jones will get the job. But – and this is Kaplan's first point – it can be questioned whether historically influential philosophers were fallibilists regarding justification. Descartes, for one, is usually taken rather to subscribe to infallibilism. For him, justification, which he took to involve the clear and distinct grasping of the truth of a proposition, cannot obtain unless the proposition that is thus justified is in fact true. From Descartes' perspective, then, there could not be a Gettier problem. This is the first reason why Kaplan thinks that the Gettier problem is not as important as it is commonly taken to be; contrary to popular opinion, it does not challenge an account of knowledge that is properly called "traditional".

We will leave this aspect of Kaplan's argument aside and focus on the second element of his reasoning, which is more relevant here because it involves, at least implicitly, considerations of goal setting. Kaplan's second reason for downplaying the Gettier problem stems from his particular account of what it takes for a philosophical problem to be significant. Surveying some historically important debates, such as the logical positivists' concern with a verificational theory of meaning, he concludes that what makes a problem important is the extent to which solving it "succeeded in advancing or clarifying the state of the art of inquiry" (p. 354). But it is not clear, he continues, how solving the Gettier problem would yield insights in this regard.

Kaplan invites us to consider, as a preliminary, the following issue for the so-called traditional account: Suppose that you as a responsible inquirer have considered all the available evidence, whence you conclude that p. Having done so, there appears to be no further point in asking whether you also *know* that p. Once you have satisfied yourself that you have a justified belief, you have thereby also satisfied yourself that your belief is true:

> From where you sit, determining whether you believe p with justification and determining whether you know that p comes to the same thing. But then, far from being integral to your pursuit of inquiry, distinguishing the propositions you know from those you don't know is, on the justified-true-belief analysis, a fifth wheel. (p. 355, notation adapted)

Kaplan proceeds to argue that the same issue arises for the post-Gettier account of knowledge:

> In so far as you are satisfied that your belief in p is well founded, you will ipso facto be satisfied that you have not inferred p from a false premise – otherwise you would *not* think you had good reason for concluding that p. Just as on the justified-true-belief analysis, determining whether you believe p with good reason and determining whether you know that p come to the same thing. (p. 355, notation adapted)

Kaplan's conclusion is that "what you know doesn't matter" (p. 362).

Kaplan's argument can be reconstructed as one based on considerations of goal rationality. From this perspective, the first part – concerning the traditional concept of knowledge – is practically identical to Rorty's (later) argument to the conclusion that truth is not a goal of inquiry; only justification is. Kaplan is, in effect, appealing to what I have called Rorty's Principle: if (i) two goals G_1 and G_2 are end-state evaluation equivalent for a subject S, and (ii) G_2 is more directionally complete than G_1, then G_2 is more rational than G_1 for S.[8] We saw that this principle is not plausible from a goal theoretical perspective because it fails to take into account the

---

[8] Since Kaplan's paper appeared 10 years before Rorty's, calling the principle in question "Kaplan's principle" would do more justice to the actual chronology. However, while Kaplan's argumentation is on the whole clearer than Rorty's, it must be said, to his credit, that Rorty is more explicit about the particular fact that he is addressing a problem of rational goal setting. Also, as I mentioned, Rorty's 1995 paper reflects ideas that he has expressed in earlier works, including works from the 1980s.

motivational aspect of visionary goals such as the goal of truth. Since the second part, about Gettier, presupposes the first part, it is just as uncompelling.

One could add that while the goal of attaining a justified belief where the justification is not based on false premises may seem to coincide with the goal of attaining a justified belief *simpliciter* because, in Kaplan's words, "none of us needs to be taught that an argument with a true conclusion does not carry conviction if that conclusion rests upon a false premise" (p. 359), there are circumstances in which the former may have some action-guiding qualities that the latter lacks. For instance, if the inquirer is investigating an emotionally loaded issue, in which she has strong personal stakes, a reminder that her position is not stronger than her weakest premise may prevent her from wishfully adopting premises that support her favored conclusion.

7 Sartwell on Knowledge as Mere True Belief

In his paper from 1992, Crispin Sartwell seeks to establish the, for a traditional epistemologist, surprising thesis that knowledge is mere true belief. His argument proceeds from the premise that knowledge is the overarching or ultimate goal of inquiry. What we seek in inquiry is, above all, knowledge. For why else, he asks, should knowledge occupy such a central place in epistemology?

Now suppose that knowledge, *pace* Kaplan, is justified true belief. There are two ways of conceiving the value of justification. Either it derives wholly from the value of true belief or it does not do so. Consider the first possibility. In that case, justification is merely a criterion of some part of the goal of inquiry, namely true belief, and cannot therefore itself be part of that goal, or so Sartwell believes. Consider instead the other possibility, i.e. that of justification being valuable independently of truth. In that case, Sartwell thinks, knowledge is an incoherent concept combining two independently valuable components which cannot always be realized simultaneously. But knowledge is not an incoherent concept. Hence, knowledge is mere true belief.

Sartwell is here assuming, in the first part of his argument, that an ultimate goal cannot contain parts which are valuable only as criteria of other parts of that goal. That is the reason why he thinks justification, if it is merely a criterion of truth, cannot be part of knowledge, if knowledge is conceived as involving truth and as being at the same time the ultimate goal of inquiry. What can be said of this part of the argument from the point of view of the general theory of goal-setting? It is indeed in the spirit of SMART+ to separate ultimate from instrumental goals in the description of a goal system, although as far as I can see, that framework does not strictly speaking disallow ultimate goals that have instrumental goals as parts.

It is rather in the second part of his argument that Sartwell may have failed more substantially. Consider his argument for thinking that if justification is valuable independently of its relation to truth, then knowledge, in the sense of justified true belief, is an incoherent and therefore useless notion. To be specific, he levels the following objection against William Lycan's proposal that justification has certain independent virtues related to explanatory elegance:

> [K]nowledge turns out to be (at least) true belief that is generated by adaptive explanatory techniques. But this seems odd: now that we recognize two primitive epistemic values, they may well conflict. For example, is it good to believe, in some circumstances, highly explanatory falsehoods? The account surely leaves some such cases strictly undecidable, since it describes both elegance and truth as intrinsic values. But is this plausible? Surely, we might want to say, though it can be useful to believe all sorts of falsehoods, it is always epistemically good to believe the truth. It may be useful, for example, for me to have cognitive technique that causes me to believe that I have all sorts of positive qualities to an extremely high degree ... But it is not a good thing epistemically to believe such things if they are false.

And he proceeds: "But we cannot coherently demand that we follow both of these as ultimate aims, because they may and in fact will conflict. Then we are left with an internally incoherent concept of knowledge."

The critical general issue here is whether the fact that two goals may sometimes conflict shows that they cannot both be ultimate ends. Sartwell thinks that this is indeed the case. However, as we saw earlier there are numerous examples of goal systems in management or technical contexts listing as ultimate aims goals that can, or even can be expected to, conflict. To take an example, the overall goal of the Swedish transport system is in fact a combination of two goals: economic efficiency and long-term sustainability. It is quite easy to imagine circumstances in which both goals cannot be attained at once. We need only imagine a case in which the economically most efficient system, because of the natural resources it consumes, is one which can only operate for a relatively short period of time. We recall that environmental politics is another area in which potential goal conflicts abound.

What follows for the purposes of Sartwell's argument? First of all, there is nothing wrong per se in having more than one ultimate goal that may conflict, such as the two goals of attaining the truth and attaining beliefs of high explanatory value. But since goal conflicts can be expected to occur, it would make sense to add a meta-rule specifying what to do when there is a tension. Such a rule could for instance stipulate that the goal of truth is to take priority in such cases over the goal of explanatory value.

The bottom line is that a proponent of the justified true belief account of knowledge can still hold that knowledge is the goal of inquiry, not least if she is willing to tie justification not to truth but to some other value independent of truth, or alternatively conceive of justification as itself intrinsically valuable. I am not recommending this move but only pointing out that Sartwell is unsuccessful in ruling it out, and – importantly – that this is so for reasons that have to do with general principles of goal rationality. In any case, taking this path may well be combined with an effort to devise a precautionary strategy for handling expected goal conflicts.

8 Conclusion

The question raised in this paper was to what extent goal setting rationality in technology is different from such rationality in science. My thesis, or conjecture, was that there is no substantial difference. The principles underlying the SMART+ theory are just as plausible for scientific goal setting as they are for technological goal setting. Moreover, the systematic advantages in viewing SMART+ as a unified theory of rational goal setting, covering all such activity, are substantial, not least for the purposes of epistemology. I tried to mount support for the latter claim by reconstructing four related epistemological debates concerning the proper goal of inquiry in the light of this unification thesis. My thesis was that all four debates depend on principles of rational goal setting that are not, or do not appear to be, generally valid. The most common oversight, from this perspective, is the failure in epistemology to take into account the motivational aspect of visionary goals, most prominently, the goal to attain objective truth. Curiously, this failure seems deeply rooted in pragmatist writings – we saw it in the writings of both Peirce and Rorty – without its apparent incompatibility with other features of pragmatism being clearly brought to the fore. I am thinking obviously of the pragmatist claim that what matters in philosophy is what makes a practical difference, from which it is concluded – to make a long story short – that truth cannot be a goal of inquiry. But the fact of the matter is that the goal of truth should rather be cherished by pragmatists as a goal which, due to its tendency to move inquirers to increase their mental effort, is as practice-affecting as one could have wished.

What also comes out, potentially, of this study is that the epistemological focus on identifying *the* goal of inquiry appears unmotivated and even somewhat obsessive. From the current perspective, there are reasons to focus less on finding a unique goal of inquiry and more on finding a plausible *system* of such goals. Finally, there is a tendency in the epistemological literature to think that potential goal conflicts should be avoided at all costs, as evidenced by Sartwell's article. But on the more general picture of goal rationality advocated here, potential goal conflicts are a fact of life, and the only conclusion that follows from the fact that goal conflicts can be expected to occur is the practical one that the designer of the goal system may want to invest some resources in planning ahead for the various contingencies that may materialize in the future.

References

Bovens, L., & Olsson, E. J. (2002). Believing more, risking less: On coherence, truth and non-trivial extensions. *Erkenntnis, 57*, 137–150.

Dieter, G. E. (1983). *Engineering design: A materials and processing approach*. New York: McGraw-Hill.

Dym, C. L., & Little, P. (2004). *Engineering design: A project-based introduction*. New York: Wiley.

Edvardsson, K. (2004). Using goals in environmental management: The Swedish system of environmental objectives. *Environmental Management, 34*(2), 170–180.

Edvardsson, K., & Hansson, S. O. (2005). When is a goal rational? *Social Choice and Welfare, 24*, 343–361.

Gettier, E. (1963). Is justified true belief knowledge? *Analysis, 23*, 121–123.

Goldman, A. I. (1976). Discrimination and perceptual knowledge. *The Journal of Philosophy, 73*, 771–791.

Houkes, W. (2009). The nature of technological knowledge. In A. Meijers (Ed.), *Philosophy of technology and engineering sciences* (Handbook of the philosophy of science, Vol. 9, pp. 309–350). Amsterdam: Elsevier.

Hughes, J. (2009). Practical reasoning and engineering. In A. Meijers (Ed.), *Philosophy of technology and engineering sciences* (Handbook of the philosophy of science, Vol. 9, pp. 375–402). Amsterdam: Elsevier.

Jarvie, I. C. (1972). Technology and the structure of knowledge. In C. Mitcham & R. C. Mackey (Eds.), *Philosophy and technology. Readings in the philosophical problems of technology* (pp. 54–61). New York: The Free Press.

Kaplan, M. (1985). It's not what you know that counts. *The Journal of Philosophy, 82*(7), 350–363.

Levi, I. (1967). *Gambling with truth: An essay on induction and the aims of science*. London: Routledge and Kegan Paul.

Mali, P. (1972). *Managing by objectives: An operating guide to faster and more profitable results*. New York: Wiley.

Peirce, C. S. (1955). The fixation of belief. In J. Buchler (Ed.), *Philosophical writings of Peirce* (pp. 5–22). New York: Dover Publications. (First published in Popular Science Monthly, 1877)

Rorty, R. (1986). Pragmatism, Davidson and truth. In E. LePore (Ed.), *Truth and interpretation: Perspective on the philosophy of Donald Davidson* (pp. 333–368). Oxford: Basil Blackwell.

Rorty, R. (1995). Is truth a goal of inquiry? Davidson vs. Wright. *The Philosophical Quarterly, 45*(180), 281–300.

Rosencrantz, H. K. (2008). Properties of goal systems: Consistency, conflict, and coherence. *Studia Logica, 89*, 37–58.

Rosencrantz, H. K., Edvardsson, K., & Hansson, S. O. (2007). Vision zero – Is it irrational? *Transportation Research Part A: Policy and Practice, 41*(6), 559–567.

Sartwell, C. (1992). Why knowledge is merely true belief. *The Journal of Philosophy, 89*(4), 167–180.

Shope, R. K. (2002). Conditions and analyses of knowing. In P. K. Moser (Ed.), *The Oxford handbook of epistemology* (pp. 25–70). Oxford: Oxford University Press.

Simon, H. A. (1956). Rational choice and the structure of the environment. *Psychological Review, 63*(2), 129–138.

Skolimowski, H. (1972). The structure of thinking in technology. In C. Mitcham & R. C. Mackey (Eds.), *Philosophy and technology: Readings in the philosophical problems of technology* (pp. 42–49). New York: The Free Press.

Chapter 10
Reflections on Rational Goals in Science and Technology; A Comment on Olsson

Peter Kroes

Abstract In the first part of my comments on Olsson I argue that the question whether or not true knowledge may be reduced to useful knowledge is not relevant for the question whether the goal of science is the same as the goal of technology. The reason is that technology is not primarily an epistemic enterprise. The goal of technology is roughly the making of useful things and the development of useful knowledge is a means to achieve this goal, not the goal of technology itself. Because what is useful is context dependent, the goal of technology is intrinsically context dependent in contrast to the goal of science. I argue that this difference in context-dependency has direct impact on when and how issues about rational goal setting in science and technology present themselves. In the second part I address the issue how the theory for rational goal setting discussed by Olsson relates to the widespread idea that rationality is only operative in the domain of means and not of goals. I argue that this theory, which stems from the field of management, is substance dependent and therefore cannot simply be transferred to science. Finally, Olsson argues that in science it may be more rational to go for the more motivational goal of true belief and not just belief, because the more motivational goal may be more achievement-inducing. I briefly point out that setting highly motivating goals may have serious drawbacks and therefore may not always be rational.

1 Introduction

Olsson compares the rationality of goal setting in science with the rationality of goal setting in management and technology and his main thesis is that "the theory of goal rationality in management and technology can be profitably transferred to the scientific domain." To support this thesis he analyses four debates about the proper goal of scientific inquiry. These debates concern questions about whether belief or true belief (Peirce), true belief or justified belief (Rorty), justified belief or knowledge (Kaplan) and knowledge as mere true belief (Sartwell) may be

---

P. Kroes (✉)
Department of Technology, Policy and Management, Section of Philosophy,
Delft University of Technology, Delft, The Netherlands
e-mail: p.a.kroes@tudelft.nl

considered the proper goal of science. He criticises the conclusions drawn about the goal of science in these debates on the ground that the principles of goal rationality upon which they are based are not generally valid. For instance, Peirce's argument that belief and not true belief is the aim of scientific inquiry is based on the following principle, that he calls Peirce's Principle: if two goals G1 and G2 are end-state evaluation equivalent for a subject S, and G1 is logically stronger than G2, then goal G2 is more rational than G1. Take for G1 the goal of true belief and for G2 belief *simpliciter*. Peirce argues that it is not possible to distinguish one goal from the other and therefore concludes that belief, the logically weaker goal, is the proper goal of inquiry. According to Olsson this principle is not generally valid because it does not take into account the motivational force of goals and therefore it cannot be concluded that G2 is the more rational goal. It may well be that the goal of true belief may be more rational because it may be more achievement-inducing due to its motivational force. In a similar way, by appealing to principles of rational goal setting, Olsson analyses the other debates and questions the conclusions drawn about the goal of science. He also concludes that application of the theory of rational goal setting to science puts into question the pervasive idea in epistemology that there is one, unique goal of science. Instead, it may be more plausible to start from the idea of a system of goals that may contain potentially conflicting goals.

Discussions about the goal of science have a long tradition and much has been written about it by scientists and philosophers of science alike. But as Olsson rightly remarks, the study of goal *rationality* in science, in comparison to the study of goal rationality in management and technology, is a "surprisingly underdeveloped intellectual territory." Olsson's paper is to be commended for venturing into this territory and his applications of the principles of rational goal setting to some debates in epistemology are of great interest because they put these debates into a new perspective (or may even be of help, as he claims, in resolving some of them). But Olsson's paper also raises a number of rather fundamental issues – and it is also to be commended for doing so – about on the one hand the goal of science and technology and about the notion of rationality on the other. In the first part of my comments I will dwell extensively on Olsson's view on the goal of technology, because in my opinion this view has direct consequences for the way he analyses the (rational) goals of science and technology. In the second part, I will raise some questions about what kind of rationality is employed in the theory about rational goal setting in technology and about Olsson's application of this theory to the goal(s) of science.

2 A Critique of Knowledge as the Goal of Technology

By way of introduction Olsson briefly discusses the goal of science and the goal of technology. He observes that there is a widely shared intuition that there is a difference in ultimate goals: whereas science strives after objective truth, technology strives after practical usefulness. He points out that whether or not this implies that there is really a significant difference in goals depends on the question whether scientific truth is different from practical usefulness. From an instrumentalist point of view, he says, this difference may be questioned because of its "pragmatic" interpretation of scientific truth. Even a realist interpretation of truth may be compatible with the idea that theory choice in science is primarily guided by the usefulness of theories for solving practical or empirical/theoretical problems. In both cases, the intuition that there is a difference in ultimate goals becomes problematic.

In my opinion there is something deeply worrying about the way he characterizes the goal of technology. In his *Introduction* Olsson appears to accept the idea that practical usefulness is the goal of technology. From his discussion it transpires that this practical usefulness is mainly to be interpreted as the practical usefulness of the *knowledge* (theories) produced in technology. He is not alone in this interpretation; it is made time and again in the literature. It underlies most discussions about whether or not technology is a form of applied science. Whether one agrees or disagrees with the idea of technology-as-applied-science, technology is taken to be primarily a knowledge generating activity with practically useful theories as its main goal. In other words, technology, just as science, is taken to be an epistemic endeavour. This is, what I would like to call, an *epistemologically biased* picture of technology. Such a picture may be true for the technological (engineering) *sciences*, but that is certainly not true for technology, more in particular engineering. The typical goal and outcome of a technical/engineering project is not knowledge, but a technical artefact or process, like a bridge, a mobile phone, an airplane, a new chemical plant, a coffee machine et cetera. In other words, the making of devices that are practically useful or bringing about desirable states of affairs in the world is the primary goal of technology, not producing knowledge that is practically useful. Of course, the making of (innovative) technical artefacts may involve the generation of new knowledge, know-how and skills of various kinds but this knowledge production is instrumental to the goal of making useful things. If indeed the ultimate goal of technology is to produce practically useful things and not practically useful knowledge (about how to produce practically useful things), then even an interpretation of truth that reduces it to mere practical usefulness does not imply that the goals of science and technology are the same, since the goal of technology is not of an epistemic nature.

It is important to stress this difference in kind of the goals of science and technology, for it puts into question some of the (tentative) conclusions of Olsson's analysis or the way he arrives at those conclusions. He argues for instance that his analysis of goal rationality in science and technology "will shed doubt on some pragmatist proposals" according to which science and technology share the same goal. The reasons for these doubts are related to the motivating force of the notion of truth because of which it may be more rational in science to aim for true knowledge than only for useful knowledge. Given what was said above, I agree with his conclusion about the goals of science and technology being different, but not with the argument upon which it is based. His argument pivots around the assumption that practically useful knowledge is the goal of technology. In my opinion, this assumption mistakes one of the means of technology or engineering for its ultimate goal.

Given that Olsson appears to endorse, in his *Introduction*, an interpretation of the goal of technology in terms of useful knowledge one would expect that he supports this interpretation with examples of goals of technology that refer to the production of useful knowledge. Interestingly, that is not the case. None of the examples of goals mentioned in Sect. 2, entitled *Goal rationality in technology*, refers to useful knowledge. In discussing issues about goal setting in general he mentions examples like the goal of becoming rich and of going to Geneva and with regard to goals in technology he mentions the design of a mousetrap.[1] These examples are much more in line with the idea that goal setting in technology is not about generating useful knowledge, but about making useful things or bringing about desirable states in the world. Of course, useful knowledge about how to realize practical (technical) goals may be crucial in guiding and motivating practical action, but as such it is a means to the goal of practical action, not the goal itself. The generation of useful knowledge about how to build a bridge may be an important step in achieving the goal of building a bridge, but the goal of generating this useful knowledge about bridge building is not to be confused with the goal of building a bridge or the usefulness of a bridge.

My disagreement with Olsson about the goal of technology does not by itself undermine his claim that the rationality of goal setting in technology may be fruitfully transferred to science. My aim so far has been merely to point out that in my opinion it is a mistake to ascribe to technology the goal of the production of useful knowledge. By doing so the goal of technology is put in the same domain as the goal of science, namely in the epistemic domain, and consequently the obvious question of whether or not these epistemic goals are the same presents itself. It remains to be seen whether the ascription of a goal to technology that is different in kind from the goal of science has consequences for the question whether the rationality of setting goals in both domains is similar or not.

3 The Context Dependency of the Goal of Technology

On the face of it there appears to be a difference in the way goals are set in science and technology that originates from or is related to this difference in kind of their goals and this difference in the way goals are set may affect the *rationality* of

---

[1] Here it might be objected that the example of a design of a mousetrap is an example that confirms Olsson's claim that the goal of technology is useful knowledge, since he is not referring to the making of a mousetrap. However, it is clear that he has the latter in mind, since he writes that the product of a mousetrap designer "should not harm or restrict the freedom of the mouse in any way"; clearly, it does not make sense to claim that useful knowledge (in the form of a design of a mousetrap) as such restricts the freedom of a mouse in any way.

goal setting in both domains. To clarify this difference, let us follow Olsson in his conjecture that the principles underlying the rational theory of goal setting "are just as plausible for scientific goal setting as they are for technological goal setting" and let us assume that by applying these principles to science we arrive rationally at some plausible system of goals for science. As Olsson points out, this system of goals may contain goals that are potentially in conflict; if conflicts arise scientists will have to deal with them in concrete situations. This system of goals then applies to any scientific research project; it is, so to speak, the fixed, overarching end of any scientific inquiry in so far that inquiry is a (purely) epistemic endeavour. In principle, this immutable end determines the rationality of theory choice in science: whenever scientists are confronted with competing theories they ought to opt rationally for the one that brings them closer to this immutable end. In case the immutable end contains conflicting goals problems may arise about what it means for one theory to be closer to this end than another.[2]

If we apply this view on the rational goal setting of science to for instance the study of the physical phenomenon of the rainbow, the following picture emerges. All research projects on rainbows, *qua* scientific research projects, share the same ultimate goal (depending on what is the outcome of the rational goal setting for scientific inquiry the following goals may play a role: to understand this phenomenon, to find out its 'true' causes, to predict the occurrence of this phenomenon et cetera). Once the ultimate system of goals of science in general has been fixed, it can be straightforwardly applied to this specific case and no further (re-)setting on rational grounds of the end of inquiry prompted by specific features of the phenomenon of rainbows is necessary; more in particular, no rational resetting of the end of inquiry on the basis of the outcomes of inquiries into rainbows over time is necessary.[3]

Now, if we shift focus from science to technology the following question arises: What would be the technological counterpart of this immutable scientific end? Given our discussion about the goal of science and technology above, the most obvious candidate is the making of useful things.[4] At first sight this immutable end of technology can be applied straightforwardly to specific cases of engineering, for instance the design and making of a useful car or a useful mobile phone. On closer

---

[2]In the following I will ignore the fact that the system of goals of science may contain conflicting goals. If we take that into account, the following analysis may become somewhat more complicated because the idea that there is one ultimate goal for science may become problematic. However, the main point concerning the difference in context dependency between the goal (system of goals) of science and technology remains valid.

[3]Of course, over time changes in which features of rainbows were considered to be the most important to study and explain may have occurred; however, that does not imply a resetting of the goal of this research as described above.

[4]According to many engineering codes of conduct the paramount goal of engineering/technology is to serve the public. For present purposes I will assume that this goal is more or less taken account of by the notion 'useful' in the expression of useful things.

look, however, a different picture of the goal of and goal setting in engineering projects from the one in scientific research projects emerges.

Let us have a closer look at goal setting in engineering projects. It may happen that within a specific engineering project the goal set at the beginning of the project, for instance to design a useful car or mobile phone satisfying a given list of specifications, may be achieved without having to readjust the goal during the project. No issues that necessitate a resetting of the goal of that particular engineering project come up and so the goal remains fixed during the project. However, this situation may be more the exception than the rule. Often during an engineering project the goal of the project has to be readjusted on the fly; this may be the case for various reasons arising from developments within the engineering project or its context: because the given list of specifications cannot be satisfied given the available resources, because of changes in the available resources, because of conflicts between various specifications et cetera. Then, a (rational) resetting of the goal of the project is called for. If there are conflicts between certain specifications trade-offs between those specifications will have to be made. What happens in these cases is that the goal of a useful car, which was originally defined in terms of a particular list of specifications, is redefined and so the goal of the engineering project is redefined.

Prima facie, the situation with regard to goal setting in specific scientific research projects may look not very much different from the one sketched above with regard to engineering projects. Also in scientific projects it may turn out to be necessary to redefine the original goals that the project set out with because of unexpected research outcomes or because of contextual developments. Thus, rationality issues about goal setting within research projects in the sense of which research questions to pursue, which experiments to perform et cetera, may come up. However, these rationality issues touch upon the specific goals of scientific research projects, not upon the ultimate end of scientific research. No resetting of that end on the basis of the outcomes of particular research projects appears necessary. All scientific research projects, whatever their specific project goals, appear to share a common, ultimate goal.

It is precisely the idea of a shared ultimate goal that is problematic when it comes to engineering projects. To see why, let us shift attention from individual engineering projects to series of consecutive engineering projects intended to improve on previous versions of a product. Over the years the criteria for usefulness of cars and mobile phones have shifted considerably which is reflected in changes in the lists of specifications for cars and mobile phones. Because of these changes the goals of engineering projects for designing and making cars and mobile phones are also continually changing. Going from one engineering project to the next a resetting of goals occurs and issues about doing this rationally turn up. Various kinds of factors play a role in (re)setting these goals, ranging from market position, legal constraints, consumer wishes, production facilities, financial position et cetera. This diversity of relevant factors may make goal (re)setting in these situations a very complicated matter.

Now it might look as if, similar to the shared overarching goal of all scientific research projects on rainbows, there is likewise a shared overarching goal for all car or mobile phone making projects, namely the design and making of the most useful, best or optimal ('true') car or mobile phone. One argument to support this idea is that from a technical point of view it has proven to be possible to produce in the course of time increasingly better and technically more sophisticated cars and mobile phones. This suggests that there might be something like a technically best, and therefore most useful, design for these technical artefacts. However, in contrast to the science of rainbows there is no such overarching, Archimedean goal in technology. First of all, the notion of the technically best or optimal car or phone does not make sense. Without going into details, one of the reasons for this is that there is no single criterion on which to measure the overall technical performance of cars or mobile phones. Their lists of specifications, even if all strictly non-technical specifications are filtered out, usually contain a number of different criteria on which their performance is to be assessed. It has been shown that under very general conditions it is not possible to aggregate these different criteria into an overall performance criterion on which to assess the technical performance of different design options (Franssen 2005). In such situations, notions like the technically best or optimal car or mobile phone do not make sense and with it the idea that there might be a gradual approach to their realization by successively resetting the goals of engineering projects such that they will bring us closer to these technically 'ideal' objects.[5] That is the reason why the notion of technical improvement should be used with care.[6] Secondly and more importantly, even if the notion of a technically best object would make sense, the inference from technically best to most useful is very problematic. Usefulness is a context dependent notion; what may be (most) useful in one context, may not be so in another. This means that if the making of the best technical artefacts in the sense of most useful ones is taken to be the ultimate goal of technology, this goal is highly context dependent. All in all, as Thomas Hughes, one of the leading historians of technology remarks "From Hunter's monograph historians and students learn about the realities of technology transfer and the absurdity of arguing that there is one best engineering solution." (Hughes 1991, p. 16).

The upshot of the foregoing is that it points to a significant difference in the setting of goals in science and technology. Once the overall goal (system of goals) of science has been set, be it theories that are true or empirically adequate et cetera, no resetting of that goal in specific research projects on the basis of the outcome of those research projects appears necessary. A similar overarching goal for engineering projects is lacking. It may be agreed that making (optimally) useful

---

[5] A similar kind of reasoning may apply to science in case its system of goals involves various independent criteria for measuring the performance of theories; see, for instance, Zwart and Franssen (2007) who argue that this is the case if a notion of verisimilitude that places content and likeness considerations on the same level is taken as the goal of science.

[6] This does not exclude the possibility of Pareto improvements.

artefacts is the overall, ultimate goal of technology, however, in contrast to the situation in science, this goal is not of much help in fixing (constraints on) the goal of a particular engineering project: How to derive specific lists of specifications for engineering projects from the goal of making a useful car or useful mobile phone? For each individual engineering project as well as in series of engineering projects goals have to be set and reset or readjusted on the basis of 'historically contingent' considerations about what is considered to be useful. There is no predefined goal for a particular engineering project – the technically best, let alone most useful, car or mobile phone – that so to speak by default drops out of the ultimate goal of technology. The reason for this is that usefulness of technical artefacts is a thoroughly context dependent notion and because of this contextual factors play a primary role in the rationality of goal setting in engineering projects.

This difference in (the rationality of) goal setting in particular research and engineering projects reflects a rather fundamental difference in what science and technology in my opinion are about. In its core, science is about creating abstract epistemic artefacts whose aim is to faithfully or reliably represent a pre-given world; this aim remains unaffected whatever aspect of the world is studied and once the criteria for a faithful or reliable representation have been set they apply indiscriminately to any scientific research project. *Prima facie* a similar kind of reasoning may be given for technology, namely technology is about the creation of (physical/material) artefacts for usefully intervening in the world. Whatever kind of technical artefact is being created, this aim remains the same, analogous to the case of science. However, with regard to (the rationality of) setting goals there appears to be a crucial difference between creating abstract artefacts that faithfully or reliably represent the world and physical artefacts that are useful for acting in (changing) the world. What it means for representations to be faithful or reliable and the epistemic criteria for assessing their faithfulness or reliability are taken to be context independent, whereas what it means for a technical artefact to be useful is a historically contingent matter and the criteria for assessing their usefulness are intrinsically context dependent. In case the epistemic criteria for faithful or reliable representations would be as context dependent as the criteria for practical usefulness, there would be a real danger of a thorough (cultural) relativism of scientific knowledge.

It may be objected that this difference in context dependency becomes less pronounced or even disappears when in a pragmatist or instrumental spirit the goal of science is taken to be the creation of abstract artefacts (such as theories) that are useful. However, much depends upon how the notion of usefulness is interpreted. It may be interpreted in an epistemic or in a practical (technical) sense and the two are not to be confused. If it is taken in an epistemic sense, then a strong case can be made that the above difference still remains. To illustrate this, take the following quote from Olsson with regard to Peirce's pragmatism:

> We recall that, for Peirce, belief or opinion is, by definition, that upon which an inquirer is prepared to act. Hence, Peirce is proposing to reduce the goal of scientific inquiry to the goal of attaining that upon which we are prepared to act. Peirce's enterprise could be interpreted as presenting the TU-intuition [the intuition that science has truth as goal and technology

usefulness; PK] with a direct challenge, especially if the kind of action that is referred to is practical action, a reading which is not alien to Peirce's pragmatism.

For Peirce the goal of scientific inquiry is the "fixation of belief"; this is an epistemic goal: it is the production of useful knowledge in the sense of beliefs upon which we are prepared to act (also in practical matters). This usefulness of a belief (knowledge), however, is not to be confused, as Olsson appears to do, with the usefulness of a practical action that we are prepared to perform on the basis of it. We are more or less back to our earlier example of the difference between the goal of producing useful knowledge about building bridges (on the basis of which knowledge we are prepared to act) and the goal of building bridges and the usefulness of bridges. Even if the goal of scientific inquiry is interpreted in a pragmatic way as some form of useful knowledge, this goal may be largely context independent. In scientific inquiry the goal may be useful knowledge in the sense of the fixation of beliefs on the basis of which we are prepared to act no matter what our particular practical goals are. This is very much in line with Peirce's idea that truth is what the community of inquirers converges upon in the long run. This convergence in useful beliefs, beliefs upon we are prepared to act, does not imply a convergence in practical (technical) usefulness of the actions (or their outcomes) based upon these useful beliefs. In the field of practical usefulness it is much more difficult to make sense of the notion of a convergence of usefulness; again, there is no convergence to something like the most useful car or mobile phone.

To conclude this part of my comments, I have argued that the question whether or not true knowledge may be reduced to useful knowledge is not relevant for the question whether the goal of science is the same as the goal of technology. Furthermore, if the making of useful things is taken to be the goal of technology, then the goal of technology is intrinsically context dependent in contrast to the goal of science and this difference in context-dependency has direct impact on when and how issues about rational goal setting in technology present themselves.

4 What Kind of Rationality?

I now turn to Olsson's analysis of the *rationality* of goal setting in science and technology. As already pointed out, Olsson's thesis is that a particular theory of rational goal setting of technology and management can be fruitfully transposed to science. This theory of goal setting is known under the acronym SMART: rational goals should be (S)pecific, (M)easurable, (A)chievable or (A)ccepted, (R)ealistic and (T)ime-bound. This theory has been further elaborated by Edvardsson and Hansson into a framework that Olsson refers to as the SMART+ theory of rational goal setting. One question that immediately comes to mind is what kind of rationality is involved in this goal setting. Indeed, there is a long tradition in philosophy to deny that goals may be set in a rational way. As Hume remarked "Reason is, and ought only to be, the slave of the passions...". Since the fixation of goals or ends falls outside the province of reason or rationality, he (in)famously claimed that it is not contrary to reason "to prefer the destruction of the whole world to the scratching of my finger" (Hume 1969 (1739–40), pp. 462–3). According to this tradition rationality can only do its work in the realm of choosing the right means when goals are given and therefore the only form of rationality is instrumental rationality. So considered, the notion of goal rationality appears to be rather an oxymoron, the more so since the SMART+ theory of rational goal setting is borrowed from the field of technology and management in which 'instrumental thinking' plays a dominant role.

Olsson does not address this question, but Edvardsson and Hansson do (Edvardsson and Hansson 2005). They point out that their framework for rational goal setting does not concern substantial aspects of goals, but "non-substantial, or structural, properties" of goals such as consistency of goals (p. 344). They refer to various non-substantial properties mentioned in the literature on goal setting in private and public management such as "clear, concise and unambiguous, within the competence of man, challenging, measurable, evaluable, integrative, complex, dynamic, transdisciplinary, applicable, participatory and understandable" (pp. 344–5). They propose an outline of a systematic account of these non-substantial requirements on goal setting which pivots around four non-substantial criteria for rational goals, namely, goals should be precise, evaluable, approachable and motivating.

The question I would like to raise here is whether their theory of rational goals really concerns non-substantial aspects of goals and thus may be expected to be generally applicable in any context of goal setting. To raise some initial doubts about this: is, for instance, the requirement of goal consistency really substance independent? It may not be rational for me to set the goal of going to Geneva this evening and staying at home this evening. But what about raising your children and having the goal of protecting them from harm and the goal of fostering an independent, autonomous attitude in them? As any parent may have experienced, on occasion these may be conflicting goals. Does this mean that it is irrational for me to have these conflicting goals? Let me focus on the SMART criteria. In my department these are used in my yearly review with my superior. I have always been troubled by the question whether these criteria of goal setting can be sensibly applied to scientific work. If not, then this raises serious doubts about Olsson's thesis that the theory of rational goals from management and technology may be fruitfully applied to science.

One line of reasoning that leads to doubt the non-substantial nature of the SMART criteria goes as follows. These criteria have been developed and are used by managers with certain substantive goals in mind; a prominent one among these goals is that they want to be able to control and evaluate the projects for which they are held accountable by their superiors. To that end, they see to it that agreements with project collaborators about project goals and resources are stated in a way that allows them to evaluate the outcome of projects in an uncontroversial, intersubjective ('objective') way. Here the SMART criteria do their work; they help managers and their collaborators to set goals that may be assessed in unambiguous ways (for example, in my yearly review, the outcome of my research is measured in terms of the number of papers published in ISI-journals, not in terms of whatever substantive progress I may have made in my research). In this way, these collaborators in turn can be held accountable for their work. From this perspective, the SMART criteria are simply an instrument to help managers to achieve their substantive goals, which means that they are not substance-independent, but directly related to or derived from particular goals. What gives the SMART criteria an air of being non-substantive is that they are operative at the managerial meta-level and put only very general constraints on setting particular substantive goals for projects. But the managerial level has its own substantive goals and the instruments to realize these goals, of which SMART is an instrument that helps managers to fix the goals of projects in a way that is conducive to the realization of their goals. According to this line of reasoning the kind of rationality involved in rational goal setting as discussed by Edvardsson and Hansson falls squarely within the domain of instrumental rationality and is fully compatible with the widespread and long-held idea that rationality is only operative in the field of means and not of ends.

Another way to question the non-substantial nature of the SMART criteria is to apply them to science directly. Thus, in order for the goals of a scientific research project to be rational they should be specific, measurable, achievable, realistic and time-bound. Typically, one would like to impose these criteria on the goal of a PhD research project. A PhD-student has to prove that (s)he is able to become an independent researcher and in order to assess whether that is the case, the project goal better satisfies the above criteria to a large extent. Otherwise the PhD-student might be blamed for the failure of a project, whereas in fact the thesis supervisor was to blame because of setting an unachievable goal. Note that in this case the SMART criteria are instrumental to a specific meta-level goal, namely the goal to offer a PhD-student the possibility to prove that (s)he is able to become an independent researcher. So, we are more or less back in the managerial situation described above: the SMART criteria are applied in order to set rationally a substantive meta-level goal, which results in very general constraints on setting the substantive goal of the PhD project.

What about applying the SMART criteria to the goals of research projects when there are no meta-level goals as in the case of PhD projects? Then the SMART criteria operate, so to speak, on the level of the 'intrinsic' goal of the research project. It appears that they may be applied fruitfully to routine like research projects in which the same research questions and experimental procedures are employed (for instance, the same experimental procedure for yet another sample of a particular kind of substances). But for innovative, explorative, open ended research projects it may be much more difficult to formulate goals that satisfy the SMART criteria. At the frontiers of science, not only research goals, questions and outcomes are usually heavily contested, often it is not clear whether some goals are achievable, let alone that it is possible to set time constraints on achieving certain goals. If the SMART criteria are really non-substantial constraints on rational goal setting, one would have to conclude that rational goal setting in these domains of science is not possible. Instead of going for this conclusion, I am inclined to question the claim that the SMART criteria are non-substantial. They have been developed within a specific (managerial) context for particular kinds of activities. That they are not applicable to certain kinds of scientific research projects – the kind which is usually considered to be most valuable for fostering the ultimate goal of science – may be due to the fact that the SMART criteria are to a large extent substantial in nature (or context dependent). Then the conclusion to be drawn is that the goal underlying the SMART criteria does not fit certain kinds of activity going on in science. Note that this conclusion partly undermines Olsson's main thesis that the principles of rational goal setting in management and technology can be fruitfully applied to the rational goal setting in science.

5 Conclusion: Rationality and Motivating Goals

My final point of comment concerns Olsson's claim that it may be more rational to go for the more motivational goal even if the end states of the more motivational goal are not evaluatively different from the end states of less motivational goals. That is the reason why he rejects Peirce's Principle and Peirce's conclusion that the goal of inquiry is just belief and not true belief. In his opinion epistemologists in general have failed to take into account the motivating role of visionary goals in their debates about the goal of science, in particular the goal of aiming for objective truth. In his *Conclusion* he writes:

> Curiously, this failure seems deeply rooted in pragmatist writings [...] without its apparent incompatibility with other features of pragmatism being clearly brought to the fore. I am thinking obviously of the pragmatist claim that what matters in philosophy is what makes a practical difference, from which it is concluded – to make a long story short – that truth cannot be a goal of inquiry. But the fact of the matter is that he goal of truth should rather be cherished by pragmatists as a goal which, due to its tendency to move inquirers to increase their mental effort, is as practice-affecting as one could have wished.

For Olsson there seems to be no doubt that the visionary goal of objective truth plays a positive role in science:

> Indeed, the goal of true belief, or the goal of truth for short, does sound more inspirational than the goal of settling belief. Many people, not least those equipped with a scientific mind, will go to almost any length to find the truth of the matter, sometimes even in practically insignificant affairs. Disregarding the special case of religious faith, comparatively few would be willing to incur similar personal and other costs for the sole gain of settling a corresponding opinion.

This may sometimes indeed be the case. However, in my opinion there is also another, negative side to the coin of aiming for objective truth in case objective truth cannot be distinguished from mere belief, which seriously undermines the claim that it is more rational to go for objective truth than just settling belief as the aim of inquiry. From the perspective of rational goal setting, the worry that I have is a general one, namely, that it may be questioned whether any goal that combines the properties of being highly motivating and being not (fully) end-state evaluable may be a rational goal in the sense of achievement-inducing. Objective truth as a visionary, strongly motivating goal may seduce inquirers to think that they have (partly) attained that goal. Once that is the case, they, to paraphrase Olsson, will go to almost any length to defend 'their truth' of the matter and they can do so precisely because the goal of objective truth is not end-state evaluable or not evaluatively different from less motivating goals. According to Kuhn's analysis of science we see this behaviour also in science (Kuhn 1962); some scientists stick to an old paradigm 'no matter what', whereas others 'convert' to a new paradigm. That is the negative side of the coin of aiming for highly motivating goals that are not (fully) end-state evaluable. Here, Kuhn's analogy with religious conversion is telling; scientists may become dogmatic about their own truth. In my opinion, the dangerous sting in Olsson's analysis is exposed precisely by his reference to religious faith in the last sentence of the above quotation. The detrimental effects of the strongly motivating goal of objective truth when it comes to matters of religion are too well known to be spelled out in more detail and suffice to show that, after all, it may not be rational to opt for a strongly motivating goal when that goal is not end-state evaluable or not evaluatively different from less motivating goals.

References

Edvardsson, K., & Hansson, S. O. (2005). When is a goal rational? *Social Choice and Welfare, 24*, 343–361.
Franssen, M. (2005). Arrow's theorem, multi-criteria decision problems and multi-attribute design problems in engineering design. *Research in Engineering Design, 16*, 42–56.
Hughes, T. P. (1991). From deterministic dynamos to seamless-web systems. In H. E. Sladovich (Ed.), *Engineering as a social enterprise* (pp. 7–25). Washington, DC: National Academy Press.
Hume, D. (1969 (1739–40)). *A treatise of human nature*. Harmondsworth: Penguin.
Kuhn, T. S. (1962). *The structure of scientific revolutions*. Chicago: University of Chicago Press.
Zwart, S. D., & Franssen, M. (2007). An impossibility theorem for verisimilitude. *Synthese, 158*, 75–92.

Chapter 11
The Naturalness of the Naturalistic Fallacy and the Ethics of Nanotechnology

Mauro Dorato

Abstract In the first part of this paper, I clear the ground from frequent misconceptions of the relationship between fact and value by examining some uses of the adjective "natural" in ethical controversies. Such uses bear evidence to our "natural" tendency to regard nature (considered in a descriptive sense, as the complex of physical and biological regularities) as the source of ethical norms. I then try to account for the origins of this tendency by offering three related explanations, the most important of which is evolutionistic: if any behaviour that favours our equilibrium with the environment is potentially adaptive, nothing can be more effective for this goal than developing an attitude toward the natural world that considers it as a dispenser of sacred norms that must be invariably respected. By referring to the Aristotelian notion of human flourishing illustrated in the first part of the paper, in the second I discuss as a case study some ethical problems raised by mini-chips implantable in our bodies. I conclude by defending their potential beneficial effects of such new technological instruments.

1 Introduction

Despite an increasing attentiveness to technology – focussing in particular on the epistemology of artificial models of natural systems and on the use of simulations and numerical calculations allowed by more and more powerful computers – philosophers of science still seem to be more devoted to the foundations of the so-called "pure" sciences rather than to the clarification of the conceptual connections between applied science and traditional philosophical issues. The consequence is

---

Research for this paper has been supported by the Spanish MINECO grant FFI2011-29834-C03-03. I thank Karim Bschir and Matteo Morganti for comments on a previous version of this manuscript. This paper develops some arguments contained in Dorato (2012).

M. Dorato (✉)
Department of Philosophy, Communication and Media Study, The Roma3 University,
Viale Ostiense 234, 00146 Rome, Italy
e-mail: dorato@uniroma3.it

that, at least in the last century, a philosophical reflection on the nature of technology has been left only to continental philosophers and theologians, who are typically animated by a negative attitude toward it (see for instance Franssen et al. 2010). The general public is therefore often misleadingly frightened by the cultural influence of these intellectual circles.

Traditionally, two philosophical issues have been considered to be central in the philosophy of technology. The first (Q_1), more discussed by philosophers of science, involves the question whether it is technology or pure science that is the driving force of our increased understanding of the natural world. The second (Q_2), much more discussed by continental philosophers and only recently by analytic philosophers, concerns the relationship between technology and human values in general.[1]

(Q_1) As to the former question, few remarks here will have to suffice. The first is that we are aware from historical studies on science that the role of technology has been essential both for the first scientific revolution (Rossi 1970) and for the so-called "second scientific revolution", a process that, according to Bellone (1980), took place during the second half of the nineteenth century and culminated in the birth of relativity and quantum theory in the early part of the twentieth century. Thermodynamics for example is a classical case in which an inquiry into the efficiency of the steam engine – a problem of engineering – has preceded and made possible the formulation of phenomenological principles in thermal physics and, subsequently, of theoretical laws in statistical mechanics.

The second remark is that the politically, economically and socially central problem[2] whether new discoveries in pure science precede or are preceded by applied, technology-oriented science, presupposes that the distinction between pure and applied science is clear-cut. But historical evidence shows that such a distinction is at best one of degree, and even in the discipline where it might seem to be more at home, *mathematics*, pure and applied mathematics are in constant and fruitful interaction.

As is well-known, the branches of mathematics that are regarded as *pure* often do not remain "unapplied" for long, and are sometimes those that unexpectedly display more "applicative" or "technological" power. Abstract computability theory, a branch of pure logic, has become the basis for the production of computers, and has therefore been the springboard of a good part of the world economy today. The application of number theory to cryptography is a second well-known example that,

---

[1] Today many more questions are being discussed, but here I refer just to these two traditional issues.

[2] The political, economical and social importance of the problem of the relation between pure and applied research depends obviously on the fact that many governments, in periods of economic crisis, tend to cut budgets for research programs that have no immediate applications and are regarded as "pure".

together with the application of group theory to current, physics might be sufficient to illustrate the power of pure speculation in generating new technology.[3]

On the other hand, it is from branches of applied mathematics (computer-generated proofs) that have often come the solutions to problems of pure mathematics, and the role of physics (applied mathematics) as a stimulating factor in the growth of pure mathematics hardly needs any illustration (think of Newton's invention of the calculus, of the use of statistical methods in physics and the growth of probability theory or of Dirac's delta function and the theory of distribution).

(Q_2) Under the heading "ethics of technology", I think that not only should we count the already explored relationship between epistemic values (explanation, consistency, evidential strength, etc.), served by scientific theories, and non-epistemic values (economical, social, political, etc.) called into play by technology (Dorato 2004), but also the link between technology and the controversial notion of "human flourishing". Considering that contemporary neurocognitive sciences tell us that we discover our most important values through emotions and through emotions we choose,[4] it is becoming more and more important to tackle the literature from a new angle, offered by what, for lack of a better term, I will refer to as *"our emotional attitudes toward the dyad nature/technology"*. As far as I know, the perspective offered by this angle has been neglected in the analytic philosophy of technology. And yet, the above-mentioned radically negative attitudes toward technology in general – that are typical in much of what Mitcham (1994) referred to as 'humanities philosophy of technology' – are widely shared by the public and often dictate political agendas. Such negative attitudes need to be understood more thoroughly, since they might reflect deeply rooted and possibly *innate* emotional attitudes toward nature and our place in it. Until these attitudes are better understood, the ethics and politics of technology will suffer from superficiality.

More in details, the two main theses that I will articulate in this paper are as follows:

(T_1) If we want to understand the impact of new technology on the wide public (nanotechnology in particular), we must first pay attention to our pre-theoretical, emotional attitudes toward nature. Such attitudes include the fact that we tend to refer to nature as a source of ethical norms, for reasons having to do with our evolutionary past (both biological and cultural).

(T_2) Fears of technology (in particular, of nanotechnological devices implanted in our bodies, which will be the object of a brief case-study in the second part of the paper) are mainly motivated by these attitudes.

The plan of the paper is as follows. In Sect. 2, I will show the ubiquity of attempts to deduce norms from empirical generalizations taken from the biological world. A simple analysis of some of the ways in which the words "nature"/"natural" are used will reveal this fact. The well-known, resulting confusion of the fact-norm

---

[3]Consider that quantum mechanics is, on its turn, at the basis of most of today's technology.
[4]See among others Damasio (1994, 1999), LeDoux (1998) and Gigerenzer (2008).

distinction has been denounced several times from Hume onward. However, it is important to understand that the tendency to fall prey of the naturalistic fallacy is quite "natural" on our part and needs to be discussed in the wider context of the Aristotelian notion of "human flourishing". Section 3 will provide some hypotheses to explain the possible origins of this tendency, by stressing not only its possible adaptive value but also the role of anthropomorphic projections of our mental and social setups onto the natural world. In Sect. 4, I will finally discuss the case study given by implantable chips by arguing that current and foreseeable developments of this form of nanotechnology are *not* so threatening after all, provided that we have a clearer understanding of the origin of our fears and that we exercise prudence and wisdom.

2 Some Paradigmatic Examples of Appeals to Nature as a Source of Ethical Norms[5]

Many of us have not come to terms yet with the rapidly changing image of our place in nature that the development of science and technology has fostered in the last 500 years, in particular for what concerns the relationship between facts and values in the application of technology. The following list of examples, which I present in the form of slogans in order to stress their rhetorical appeal, has the purpose to show the importance of the adjective "*natural*" in arguments trying to justify ethical and social norms. The comments that follow the list will set the theoretical framework against which I will discuss the particular case of implantable microchips.

1. Unequal distribution of resources is often justified by social-darwinists' slogans of the kind: "it is *natural* that the stronger prevails over the weaker";
2. "this action, this law, this rule, this technological device trespasses the limits of *nature*" is a frequently used appeal, based on an allegedly normative notion of "human nature";
3. "this is natural, *biological* food", is frequently used by environmentalists and movements that want to defend non-adulterated food;
4. "mammals are *naturally* carnivorous, or *naturally* polygamous", used against vegetarians or believers in monogamy;
5. "the (Italian) Republic acknowledges the rights of the family as a *natural* society founded on marriage"[6];
6. In the stoic philosophy we often read statements insisting that our individual natures are part of a universal nature, so that we should live a life *according to nature*, where "nature" refers to one's own nature and to the universal nature.

---

[5]This list was discussed already in Dorato (2012).
[6]This is my translation of the 29th article of the Italian Constitution.

Let me briefly comment on each of these uses. Note first of all that all of the uses of "natural" in the above list, and similar others that can be found in common discourse and social/political agendas, can be classified under the opposite labels of laws, actions, behaviours, etc. that are "*according to nature*" or that are "*against nature*".

1. This first slogan was originally proposed by the ancient sophists, who introduced a fundamental distinction between what is "by nature" (*physis*) and what holds by "human convention" (*nomos*). We should notice that what holds "by nature" for the sophists concerns more or less stable regularities of the natural, biological world – like "the law of the strongest." In the Platonic dialogue *Gorgias*, for instance, Callicles contrasts such regularities with the conventions of human laws, which in his opinion were created by the "weaker" to protect themselves against the "stronger". In Callicles' view, there is a radical tension between natural and human laws, and the latter are criticized because they are "against nature".[7]

Unlike Callicles, however, we do not consider *the fact* or even the *generalization* that big fishes eat smaller ones, and similar "natural" facts, as *justifications* for the validity of an ethical or a legal principle that were to grant stronger or more intelligent human beings more rights than to weaker or less able ones. Whether the Christian precept of helping the vulnerable and the needy is going against *our* nature is doubtlessly a matter of debate (see below and note 8), but it certainly amounts to a reversal of *some* widespread regularities of the biological world. It follows that our laws and ethical values, to the extent that they defend the weak and limit the strong, are "*against nature*" (at least in part, and in a descriptive sense of "nature"), but this is no reason to criticize them from the moral point of view. Unlike Callicles, we prefer our ethical, possibly conventional or culturally induced moral convictions to what happens in nature, so as to refuse to model our institutions on the relationship between predator and prey.

In a word, ethical arguments drawn from "nature", that is, from widespread biological regularities, are unsound, even if we selected examples of "altruistic", animal behaviours. In the natural *and* in the human world, in fact, there are cooperative or "sympathetic" inclinations,[8] but they coexist with predatory and aggressive instincts. These remarks also show that it is our *prior commitment* to

---

[7]"But in my opinion those who framed the laws are the weaker folk, the majority. And accordingly they frame the laws for themselves and their own advantage, and so too with their approval and censure, and to prevent the stronger who are able to overreach them from gaining the advantage over them, they frighten them by saying that to overreach others is shameful and evil, and injustice consists in seeking the advantage over others. For they are satisfied, I suppose, if being inferior they enjoy equality of status. That is the reason why seeking an advantage over the many is by convention said to be wrong and shameful, and they call it injustice. But in my view nature herself makes it plain that it is right for the better to have the advantage over the worse, the more able over the less. And both among all animals and in entire states and races of mankind it is plain that this is the case – that right is recognized to be the sovereignty and advantage of the stronger over the weaker" (Plato, Gorgias 482e).

[8]Think of all the examples of cooperation in the animal world described by de Waal (1996). For a defence of our altruistic nature, see also Sober and Wilson (1998).

certain values (cooperation *versus* selfishness), and our attempt to justify them, that guide us in selecting those biological regularities that best match them. Such appeals to regularities of the biological world, if used to maintain that certain (nano)technological devices are "against nature", misfire.

2. Nevertheless, in public discussions scientists and engineers are very often invited not to trespass the "limits of nature", or not to go "against nature". Likewise, politicians and legislators are reminded not to pass bills that would go against nature, or "human nature". However, what does "against nature", "going beyond nature", or "overcoming the limits of nature", *mean*? At a closer look, there are *two* ways of interpreting the expressions "against nature" or "beyond nature", corresponding once again to a *descriptive* and a *prescriptive* sense of "nature".

In a *descriptive* sense, events going "against nature", or that "trespass its limits", would be events that occur very rarely, or even "miracles". These, however, would not count as events breaking the laws of nature, if by laws we mean exception-less, universal regularities described by mathematical equations, or weaker generalizations of the kind "all butterflies have wings". An exception capable of breaking a law would simply refute the known laws, but obviously would not be "against nature". There is a clear sense in which physical processes *cannot* trespass the limits of, or go against, physical laws, since laws, interpreted descriptively, constituted the very concept of *physical possibility*. If a law L were falsified by an event "going against" it, we would simply say that L is not as universal as we previously thought, and has "exceptions", or is outright false. That is, we would say either that L is not a law, or that it holds only *ceteris paribus* (see Dorato 2005). In no sense can "going over the limits of nature" or "going against nature" imply violating the laws of science regarded as descriptions of natural laws.

In the other sense of "against nature", which is more relevant to my purpose, the word "nature" is interpreted *morally*, and, in the case of human beings, calls into play the realization of *our (alleged) moral essence*. "Nature" here does not refer to the individual characters or natures of distinct human beings, but to a standard of moral perfection possibly shared by all human beings *qua* human beings. In other words, "nature" in this second sense raises the question "how human beings ought to live", not the question of how they de facto live. In this second sense, technological inventions *can* go against a morally interpreted human nature, provided of course that such a notion makes sense.

Well, does it make sense? From which premises can such shared ethical norms be derived, if not from empirical regularities characterizing our biological nature? Leaving aside the hypothesis that a human life should be lived in a certain way because God created us to fulfil his preordained aims, it seems possible to invoke a traditionally Aristotelian notion of "human flourishing", which presumably bases humans' moral behaviour on our natural, moral impulses (sympathy, compassion, love or impulses that drive us toward a fulfilled life).[9] If it were defensible, this

---

[9] Philosophers have referred to a *virtue-centered* morality (MacIntyre 1984), or to the neurophysiological basis of human flourishing (Thagard 2011).

notion of a moral human nature could be invoked to criticise those technological applications that could predictably thwart its full development or its flourishing.

There seem to be at least two objections to this notion, but they can both be met. The *first* is epistemic: since we are also endowed with passions that lead us away from self-realization, how can we identify the good impulses from the bad ones, previously and independently of a moral evaluation? (see Sidgwick 1907). The reply to this objection is that only the good passions make us really *flourish*, and that we have a natural tendency toward *flourishing*, unless a bad education distorts our "nature". Cultivating genuine friendships and, devoting one's time to meaningful work, having a healthy parent-child relationship, or possessing literacy and education, are all objective goods for human beings, or part of what we mean by "flourishing", and are not just instrumental to it.

The *second* problem might consist in the vagueness of the metaphorical notions of "flourishing" or "thriving", when referred to humans. However, the meaning of these notions *can* be clarified, since one can plausibly claim that it refers either to our being absorbed in a meaningful activity (for instance, playing, or having an instructive conversation) or to our possessing certain capacities (like having literacy, or being curious and capable of feeling wonder toward the natural world) whose exercise is an end in itself. Both engaging in an activity and having or exercising a capacity are *facts* that we evaluate positively in virtue of our emotional make-up: the notion of human flourishing, if based on our common emotional nature, seem to water down the fact-value distinction.

The reply to this objection is that it doesn't go against a certain way of construing this distinction, and that the distinction itself needs to be articulated (see below). Agreed: from the fact that well-educated persons appreciate and enjoy in a special way certain activities (say, spending time with friends they love) one cannot derive an ethical imperative *per se*. In cases of this kind, one can always raise the question "why ought we to value enjoyments of that kind?" However, it does not seem too far-fetched to reply to this second objection that posing such questions is like asking "why do we enjoy enjoyment?"

Of course, one might ask how do we find out about what contributes to our flourishing and there might be disagreement on the answers one may get. Despite individual variability, I submit (with no possibility of expanding this claim here), that *courageous, generous, loyal* and *loving* actions are universally appreciated in all cultures, despite the fact that the particular way in which these virtues manifest themselves may change diachronically and synchronically, due to differences in societal roles. A courageous soldier and a courageous politician are both appreciated but in the two cases the behaviour is largely different. It is the kind of emotions that accompany those acts that are the epistemic means to recognize what really matters for us. Self-realizing activities or the possessions of certain capacities or capabilities are not instrumental to something else, but are rather *ends in themselves*. In a word, I submit that the notion of human flourishing, which entails treating persons like ends in themselves, is decisive to create a general framework for case-by-case studies about the foreseeable consequences of any technological application.

3. "This is natural, non-adulterated, biological food," is a catchphrase often used against both genetically modified organisms (OGM) and harmful pesticides. At times, however, the fanatic fans of biological food tend to forget that agriculture, even if "biological", is not wholly natural. On the contrary, it is the product of an art ("artificial" in the etymological sense of the word), since it is the result of a complicated, contingent technique that, together with the domestication of animals, has changed the history of human beings. Of course orange-trees produce oranges "naturally", but their cultivation often requires wearing and "artificial" interventions on our part (watering, pruning, or cross-fertilizing the trees). This example is another instance of our "natural" tendency to try to justify norms by bringing to bear an illusory ideal of an unadulterated, untouched nature. If "natural" food cannot be synonymous with non-artificial, we should consider it to be suggested by what we are most used to, or what we have experimented so far. On the other hand, feeding animals with hormones or antibiotics, or spraying plants with harmful pesticides, is likely to cause problems also to humans. It follows that we should not stress the opposition "natural/artificial" or "according to nature/against nature", but rather that between what is beneficial and what is harmful for our health, where the latter is not only a good in itself, but a precondition for human flourishing. The same practical attitude should prevail on the issues surrounding the GMOs, which, however, present economical complications that cannot be analyzed here.

4. The fourth case uses animals' behaviour to defend certain human choices. The fact that mammals typically eat meat, and that we are mammals, does not make a choice for or against vegetarianism *immoral*. And yet sometimes we hear discussions in which vegetarianism is condemned in the name of what is natural, of what factually most mammals do. Another instance of trying to derive norms from natural facts, one that is also used in the name of discouraging or encouraging sexual promiscuity. Choosing as example for moral discussions those pairs of mammals that show a faithful behaviour after copulation, or alternatively, indicating male mammals that are promiscuous as a standard of behaviour for human males presupposes a previous commitment to values that cannot be justified in the name of what happens in the biological world. And yet slogans of this kind continue to be appealing for many people. Why?

5. The expression the "family as a *natural* society" recently has been the subject of hot controversies in relation to the rights of gays to marry, questions on which here I cannot enter. Suffice it to say that the adjective "natural" in this case refers to one of our biological functions, namely reproduction, with all the related behaviours, namely caring for the children etc., which are regarded as pre-legal, pre-institutional, pre-social-contract facts that the Italian constitution should take into account.

The institution of marriage is then regarded as a legalization or the "institutionalisation" of our biological function of reproduction. We should also note that the fact that human beings have the ability to reproduce, does not create by itself a moral duty to reproduce: priests, nuns and other human beings choose and have chosen not to do so. Analogously, establishing whether the only kind of "family" should be formed by people of different sex – an ethical and legal principle – cannot be justified solely on the basis of *facts* having to do with our natural capacity for reproduction, but depends on some other values. And yet such naturalistic fallacies have a remarkable impact on the large public in various western countries, and it is important to ask ourselves why.

6. "Living according to nature" is an important moral recipe in stoic philosophy, which influenced deeply later cultural movements. The stoic precept is based on the idea that everything *is as it should be*, so that our failure to accept the presence of evil is simply due to our short-sighted incapacity to perceive the whole series of events in the history of the universe. From a cosmic viewpoint, our life is but a fragment of an immutable sequence of events that is permeated by an impersonal "logos" (reason) ensuring the rationality of the whole. As the quotation referred to in point 6) above shows, the highest duty of the philosopher is to get to know the cosmic order of things, and be in command of one's passions in such a way that the unavoidable is accepted as if it were an effect of our own free will.

In a word, since the natural order of the physical world is the expression of the impersonal rationality of the universe, such order also offers automatically a moral guidance. The adaptive power of this position can hardly be exaggerated: not only is it related to the above-mentioned idea that there is a human nature in the factual and moral sense and that the two are intertwined, but it also leads us naturally to the next section, which stresses the *evolutionary advantage* of regarding nature as a source of moral rules, and therefore puts forward a possible explanation of our tendency to fall prey to the naturalistic fallacy.

3 The Natural Entanglement of the Natural with the Ethical

In these six types of rather common arguments,[10] "the entanglement of facts and values" referred to by Putnam (2004) is quite evident but devoid of argumentative power; except, importantly, when it refers to a morally characterized human nature, and therefore to the above-mentioned notion of human flourishing. In the previous section we have seen that while a natural regularity cannot in general justify a juridical norm or an ethical rule, because the latter cannot be derived from the former, we have nevertheless a strong tendency to identify in nature a foundation for our ethical values. Why is this the case? Are there explanations for this natural, tendency of human beings to find a norm in the regular order of the natural world to which we adapted during many millennia? There are at least three possible answers to this important but still neglected question,[11] partly biological and partly cultural.

---

[10] I agree that they are commoner in person-in-the-street's arguments, but this is grist to my mill, because it shows their naturalness in the sense of this section.

[11] For a brilliant exception to such a neglect, see Daston (2002, p. 374): "I wish to explore how nature could ever have been endowed with moral authority and why that authority still exerts such a powerful, if covert, pull upon our modern sensibilities, despite innumerable critiques and cautions against conflating "is" and "ought," against "naturalizing" judgments that are really social

The first is that the identification of norms in the natural, biological world that tend to regard nature as a source of values, might itself be a form of natural or cultural adaptation. We live in a natural environment to which we adapted during very long intervals of time. Consequently, keeping an equilibrium between ourselves and our environmental niche tends to increase the probability of our survival. Not by chance, much of environmental ethics is justifiably based upon the importance of maintaining an equilibrium between ourselves and nature.

Put it in a nutshell, my hypothetical explanation is as follows. Since keeping an equilibrium with our natural environment plausibly involves a certain invariance or stability of the niche in which we have lived for millennia, we probably evolved a universal attitude (which manifested itself in all cultures) to regard any radical change in our relationship with the environment as a potential *threat* to our survival. Since technology in particular nowadays is certainly perceived as the cause of such a change, it is regarded as dangerous and threatening to our survival.

If any behaviour that favours the stability and invariance of the environment is potentially adaptive, nothing can be more effective for this overarching goal than developing an approach toward the natural world that considers it as a source of sacred norms that must be invariably respected. In order to preserve the equilibrium between ourselves and the external world – an equilibrium upon which our survival obviously depend – the development of human morality might have then become inextricably entangled with the regularities of the natural and biological world. I think that the reasons for the fearful suspicions that new technological devices have always generated must be found in our evolutionary past, a factor that should be kept in mind in all public debates concerning science and, in our case, (nano)technology policies.

The second, more culturally derived reason that might account for our persistent tendency to use the notion of natural regularity (laws in the *descriptive sense*) for justifying moral laws in the *prescriptive sense* derives from anthropomorphic *projections* on nature originating from our more or less recent cultural past. By this I mean to refer to a pre-scientific attitude leading us to explain the pervading existence of regularities in the physical world with an animistic attribution of "a willing soul" also to inorganic matter. This projection might be an instance of an ADD, *Agents Detection Device*, our tendency to over-attribute intentions to the unanimated world, which has been advocated by cognitive scientists to explain the origin of religious beliefs (see Csibra et al. 1999). This over-attribution has an adaptive value because the assignment of intentionality also to unanimated entities is an application of the prudential rule "better safe than sorry": a noise in the wood might be due to an animal or to a gust of wind, but the readiness to act appropriately in all circumstances entails assuming that also the wind might carry some hostile intentions toward us.

---

and political, and against anthropomorphizing "Nature," designated with a capital N and often with a feminine pronoun." See also Daston (2004) in Daston and Vidal (2004).

In a word, this second explanation stresses the fact that the cultures from which we inherited the first forms of natural religion were struck by the notable order and regularity shown by natural phenomena. Consequently, these cultures tried to explain this order in an anthropomorphic way, i.e., by postulating the existence of spiritual entities who made it possible and explainable – angels moving the planets along precisely predictable orbits, for instance, explain why such orbits are so regular and predictable. These "entities" had to be capable of *will* and *thought*, so that they could compel nature to follow a certain course, just as human legislators impose norms of social coexistence that may not be violated.

This tendency was already evident in Babylonian thought: the characteristics of the movement of the planets, which Babylonian astronomers studied with attention and skill, were interpreted "[...] by the authors of tablets who created the library of Assurbanipal [...] as dictated by the "laws" or decisions governing "heaven and earth," as pronounced by the creating god from the beginning." (Eisler 1946, pp. 232 and 288). The same author later adds that our modern notion of universal, scientific law derives "from this mythological concept [...] of decrees from heaven and from earth," and in one of his other studies, (Eisler 1929, p. 618), he highlights the importance of the social/political condition on the way nature is represented, given that the idea of the world as an ordered entity (what the Greeks called *Kosmos*) originated, in his opinion, in Babylonian social theory. These quotations hint to a *third*, possible explanation for our persisting tendency to confuse nature and norms, one that comes from an inclination to project the social political world, with its own rules and structure, onto the natural world.

Armed with this theoretical background, we are now ready to discuss the case study given by implantable micro-chips.

4 External vs. Internal Machines: Is It Still an Important Distinction?

We have been relying for a reasonably "long" time on *macroscopic*, *external* prop-ups, like glasses, walking sticks, or electronic agendas, that are extensions of our bodies and, controversially, also of our minds (Clark 2010). On the other hand, nowadays we already have macroscopic *internal* prostheses, like artificial knees, hips, cochlear implants, pace makers, or ligaments constructed out of tendons belonging to our own bodies. In the near future, however, we might end up relying also on many *microscopical* internal parts, artificially constructed or produced *via* staminal cells, as the case maybe (Clark 2003). Is this process of "hybridation" of our bodies something to be afraid of? How should we proceed?

The following quotation from Giuseppe O. Longo, professor of computer science in Trieste expresses a widely shared viewpoint: "[first] it is impossible for the biological part of the symbiotic hybrid to keep in step with the speed of the technological evolution, and this creates a deep discomfort. The second problem is the self perception of the person. Our body is the source of our personal identity . . . the unity of body and mind would be altered by fictional prosthesis that, for instance, could alter the capacity of our memory" (Longo 2008).

The first fear expressed by Longo is grounded on the radically different speeds distinguishing technological and biological changes. And this reinforces the already stressed evolutionary fact that for our well-being it is of extreme importance to keep our relationship with the environment (and with ourselves as integral part of it) as stable and constant as possible. On the other hand, the second fear is not purely science fiction. Stefano Rodotà, an Italian jurist, refers to the hybridation yielded by such fiction prosthesis as a "post-human state": "On the 12th of October 2004 the Food and Drug Administration, has authorized the use of a very small chip that can be read at a distance, called VeriChip, to be installed under the skin of the patient and containing her whole clinical story." (Rodotà 2004) The chip, as the www.verichip.com web page advertises, "is able to offer rapid, secure patient identification, helping at-risk patients to get the right treatment when needed most." The chip would help patients affected by memory losses, impaired speech or simply patients that have lost consciousness. According to the web page, further applications of "Verychip" envisage (i) a protection against "baby switching", which amount to thousands of cases per year in the United States only; (ii) the prevention of incidents related to old people affected by mental diseases that wander around and get lost, or (iii) the possibility to have a maximum security of access to houses or banks or secret archives via a radio frequency identification. As an example of this latter application, consider that in Mexico the public attorney and some of his dependents had an implant which could not only identify them when they entered a classified archive, but could also track them in case of kidnapping. Another possible use would be tracking persons under house arrest.

Rodotà concludes his article in a very dramatic tone: "in this way the subject changes her personal and social status. A subject can always be on line, and become a networked person, configured in such a way that she can emit and receive impulses that allow others to track and reconstruct physical and mental conditions, habits, movements, contacts, thereby modifying the sense and content of her autonomy" (ibid.).

There is no doubt that cases like these deserve a very careful study and evaluation, which can only be attained via a case-by-case analysis. Given what was maintained in the previous sections, however, the particularizing strategy favoured by this pragmatic attitude can be compensated by the general outlook suggested by the morally-laden notion of human nature illustrated above, suggesting to treat human beings as ends in themselves and not just as means to an end. For instance, as noted by Rodotà, the importance of protecting personal data in cases like these should be obvious in this moral setting, especially if, say, the medical data contained in the microchip became accessible to (or even alterable by) unauthorized others. This dangerous possibility, unfortunately, reinforces an attitude that is still widespread in our cultures, for reasons that have been presented above, and which regards the whole of technological evolution as *the* dehumanising force of mankind, characterized by an exploitative and rapacious approach toward nature.

In order to contrast this attitude, not only will the following *three* brief considerations try to clarify the main issues at stake but also convince the reader that we must learn to live with the extraordinary potential offered by implantable chips and in general by nanotechnologies. With the caveat, of course, that we should try to avoid superficial enthusiasms. The first argument is theoretical, and shows how much progress has been achieved in trying to explain the macro-world in terms of the micro-world. The second shows how the progress of technological miniaturization, that will probably continue, enjoys great selective advantage in a market economy. The third discusses a few cases taken from the biomedical sciences, all marked by potentially beneficial effects.

1. As a general explanatory remark, let me begin by stressing that the technological development is following (and has at the same time greatly promoted) a scientific tendency of going "inward bound" (see Pais 1988). Such a tendency has accompanied the last two centuries physics, from the postulation of molecules to atoms, to quarks of various kind and then strings or loops (if they exist). Therefore, the take-home lesson of the last two centuries physics is that the macroscopic properties of all the physical bodies at least partially depend on, and are explainable by, the microscopic ones. Clearly, the major impact that nanotechnologies will have on our future life is going to depend on our predictably increasing knowledge this *asymmetric* dependence of the macroscopic properties of big things, *human organs included*, on the microscopic ones. This first point is put forward as a non-evaluative consideration that must be taken into account.

2. Despite their different speed, biological evolution and technological evolution obey the same abstract laws of development. Namely, a reproducing mechanism generating some variations with respect to the original, and a process of selection, which leads to the extinction of biological or technological devices, as the case maybe. Clearly, the reproduction of a machine or an artefact is based on different supports, since it depends on human brains, on culture, and therefore on education and other learning processes, and the selection in question is cultural, while the reproduction of an organism relies on chemical resources (the DNA and the RNA) and natural selection. It is the difference of the relevant and selective mechanisms that explains the disparity in the speed of change of biological organisms and technological devices. Analogously to what happens in the case of biological species, however, variations in the projects of technical artefacts explain their different impact on the market and this, in turn, creates a selective process depending on many aspects, like price, dimension, pollution, etc. The advantage of a tablet over a desktop computer having the same speed and memory is so obvious that the selective process goes in the direction of miniaturization and portability. This remark explains a strong selective push toward the miniaturization of all technological devices.

3. Let me now apply the considerations of the previous sections to the question of the role that nanotechnologies might play in our societies. It seems that we have no difficulty in accepting the idea that artificially constructed hearts, or parts thereof, or dental prosthesis, or artificial breasts, or metallic knees and hips can

be inserted in a human being. In the future, however, this tendency may increase, to that a person might become a mixture of natural and artificial parts, that is, a *cyborg* made of mixed parts. However, where should we stop? In order to answer this question, I will begin by presenting three examples involving the application of future nanotechnologies in the biomedical sciences, all three of them with potentially advantageous effects, and then discuss two possible applications of chips that might alter our cognitive capacity.

3.1. Neuro-engineers are studying the possibility that microchips implanted in the brain of an epileptic patient might detect the onset of an epileptic seizure and switch it off by cooling down the involved neurons. Times are still premature, but researchers at Washington University in St. Louis, some years ago, developed a microchip that can detect an oncoming seizure. The study, published in the journal *New Scientist*, claims that it is possible to stop seizures in the brain of rats by cooling their brain cells from body temperature (about 37 °C) to around 22 °C. The process of cooling shuts off the release of neurotransmitters, thereby rendering the cells less susceptible to seizures: apparently after the treatment the cells did not suffer any injure and worked properly. Other possibilities in this research that are still under approval by the FDA are offered by microchips programmed to detect seizures and respond via electrical shocks that are supposed to interrupt them.

3.2. Secondly, the future of pharmacology can be revolutionized by the so-called individualized medicine: one could synthesize a particular gene, insert it in a certain organism, and then obtain a molecule with a certain shape and function to be used to attack a determinate target. Along the same line, there is the well-known case of regenerative medicine, with the possibility that stem cells or other similar *totipotent* cells might create new biological tissue. This is certainly a very promising and important field of bio-nanotechnology.

3.3. Thirdly, of course, focusing only on examples taken from future, beneficial applications to medicine may render my positive attitude toward "cyborgs" superficial or biased. What about so-far imaginary applications that foresee the possibility of implanting a nanocomputer in our brains that can either modify at will our mnemonic and algorithmic capacities, or augment the natural perceptive abilities? Wouldn't this cause a collapse of our identity? However, even in these so far fictional cases there are no compelling reasons to depict a catastrophic landscape.

Patients suffering from serious prosopagnosic disorders (face-blindedness), or memory losses, for example, could be helped by a chip that – let us imagine – could correct the malfunctioning of the relevant parts of their brain. After all, we write down in external artefacts (soft or hard *agendas*) the things that we have to do in order to prevent our forgetting them. Of course, there is a considerable difference between an external and an internal device, but why should the implant of an internal agenda that could be constantly updated by our voice be regarded as something appalling? In what sense would it affect our identity?

This wholly imaginary case, however, must be regarded with due care, since the possibility of being always "on line" could give other people the chance of manipulating our own wishes and desires via a direct intervention in our brains, thereby allowing the possibility of manipulating others in a much more effective way than is permitted by today's technology.

Furthermore, it must be admitted that a chip that would enable us to remember every single experience or episode of our life not only would jeopardize our identity, but would also jeopardize our social adaptation and well-functioning. In a tale of Jorge Borges *Funes el memorioso* is incapable of forgetting:

> We, in a glance, perceive three wine glasses on the table; Funes saw all the shoots, clusters, and grapes of the vine. He remembered the shapes of the clouds in the south at dawn on the 30th of April of 1882, and he could compare them in his recollection with the marbled grain in the design of a leather-bound book which he had seen only once, and with the lines in the spray which an oar raised in the Rio Negro on the eve of the battle of the Quebracho (Borges 1962, p. 85).

It is well-known that our brains work by relying on an effective system of filtering information: this is important not only to prevent them from being cluttered with useless details but also to achieve the aim at hand. Living without forgetting would be practically impossible because empirical and phenomenological findings show that remembering is selecting and reconstructing certain aspects of our experience at the expense of others that are less salient. Borges' literary case has a real counterpart in the studies of the Russian psychologist Luria (1969), who reported his clinical experience with a patient that could never complete any task, even the simplest one, because he was constantly reminded of thousands of things that were connected with his present experience. And socially he could function very poorly. In a word, without forgetting we would not be able to remember anything and therefore we could not live because, for example, we would constantly think about our future death, while during our daily life we often *forget* that we are mortal.

One could imagine that one day it could become possible to transfer the whole ocean of data available in the web in the head of each of us, just by using a powerful microchip. However, who would want *that*? We ought not to forget that already now we can have as many (externally available) data through the web as we may want. The important question is organizing them and understanding them in more economical schemes, i.e., frame them in order to construe valuable hypotheses or arrive at significant truths. This is what, for instance, discovering a law of nature is: summarizing a lot of possible observations in a single formula. [...] "science is a form of *business*. It aims, with a minimum of effort, in a minimum amount of time, and with a minimum exertion of thought, to appropriate the maximum amount of infinity and external truth for itself" (Mach 1896, p. 14). It is exactly considering facts like these that one can easily realize that transferring the whole web in an updatable chip would not serve any purpose. And I trust that people would not even try to have such a chip implanted.

3.4. The previous point has explored the cognitive rather than the *emotional* part of the possible changes introduced by chips implantable in our heads. However, what about a future chip capable of altering our emotional states, in a way not too dissimilar from the experience machine invented by the philosopher Robert Nozick in *Anarchy, State and Utopia* (Nozick 1974, pp. 42–45)? In this book Nozick imagines a machine capable of simulating perfectly all the pleasurable experiences we may dream of, with the corresponding pleasure. Suppose now that a chip could be realized in such a way that we would not be able to tell that those experiences are *not* real. So we could experience to have a dinner with the most beautiful men or women, to win the final game of Wimbledon, to cross the Pacific with a sailboat, or receive a Nobel prize for physics or peace. Given that by hypothesis all of these would not be real experiences, but simply virtual ones, how many of us would decide to have such chips implanted without the possibility of coming back to the real world? That is, would we choose to have the chip implanted and prefer it to living a real life of toils, joys and pains?

Nozick gives the following three reasons against choosing to attach to the machine, which, if well-argued, could be extended to microchips altering our emotional states:

 (i) We don't want an ice cream because we like the experience of eating one, we like the experience because we want to eat an ice cream. "It is only because we first want to do the actions that we want the experiences of doing them." (Nozick 1974, p. 43). Here the opposition is between the real action and what doing it feels like. But we could imagine that the chip gives us the impression of acting as well. Nozick's point here does not seem convincing.
 (ii) We want to be and become a certain kind of person: "Someone floating in a tank is an indeterminate blob." (ibid, p. 43). This point appears to be more effective, as it refers to our need of living a real ethical life constituted by efforts, plans and possible failures from which to learn. However, the reply of the "nano-hedonist" could be that the chip could give us the impression and feelings of living such a life. Same as in (i): Nozick's response is not wholly persuasive.
 (iii) "There is no *actual* contact with any deeper reality, though the experience of it can be simulated." (ibid, p. 43). This point raised by Nozick seems to be the crucial one. Knowing in advance (*before* our irrevocable decision), that what we will experience after having a microchip implanted in our brain has not been gained with honest toil and has just the appearance, may deprive the expected pleasurable experience from any meaning and may convince us to refuse the "pleasure implant".

Point (iii) might be regarded as insufficient to show that hedonism is not the correct theory of our behaviour, in the sense that the *only* reason that motivates our action is the search for immediate or postponed pleasure. If hedonism were correct, choosing a microchip giving us pleasurable but "unreal" experience could still be preferred by the vast majority of human beings. But Sober and Wilson (1998) convincingly argue that hedonism is not the only motivator of our behaviour. This conclusion is compatible with the fact that if someone knew to have only few days to live, and were in terrible pain, one might decide to have the chip implanted until the final moment. And it is also consistent with the fact that the chance of refusing to attach to a virtual machine tends to be greater in subjects that are sufficiently young and in good health condition.

However, even though more empirical study is needed in order to conclude that pleasure is the only motivator of our actions (hedonism), the mere fact that human beings *can* postpone the immediate satisfaction of their needs gives evidence against the correctness of hedonism so intended. Furthermore, even merely mixed answers to the questionnaire: would you decide to have implanted a non-removable chip simulating pleasurable experience show that maybe we should not worry too much about chips that in the foreseeable future could alter our emotional states, no more than we should worry now about current abuse of drugs or alcohol.

References

Bellone, E. (1980). *A world on paper: Studies on the second scientific revolution*. Cambridge: The MIT Press.
Borges, J. (1962). *Fictions* (Eng. Trans.). New York: Grove Press.
Clark, A. (2003). *Natural-born cyborg*. Oxford: Oxford University Press.
Clark, A. (2010). *Supersizing the mind. Embodiment, action, and cognitive extension*. Oxford: Oxford University Press.
Csibra, G., Gergely, G., Bíró, S., Koós, O., & Brockbank, M. (1999). Goal attribution without agency cues: The perception of 'pure reason' in infancy. *Cognition, 72*, 237–267.
Damasio, A. (1994). *Descartes' error: Emotion, reason, and the human brain*. New York: G.P. Putnam's Sons.
Damasio, A. (1999). *The feeling of what happens: Body and emotion in the making of consciousness*. New York: Harcourt Brace and Co.
Daston, L. (2002). The morality of the natural order. The power of Medea. *The Tanner lectures* (pp. 373–411). http://tannerlectures.utah.edu/lectures/documents/volume24/daston_2002.pdf
Daston, L. (2004). Attention and the values of nature in the enlightenment. In L. Daston & F. Vidal (Eds.) (2004) (pp.100–126).
Daston, L., & Vidal, F. (2004). *The moral authority of nature*. Chicago: University of Chicago Press.
de Waal, F. (1996). *Good natured: The origins of right and wrong in humans and other animals*. Cambridge, MA: Harvard University Press.
Dorato, M. (2004). Epistemic and non-epistemic values in science. In P. Machamer & G. Wolters (Eds.), *Science, values and objectivity* (pp. 53–77). Pittsburgh/Konstanz: University of Pittsburgh Press.
Dorato, M. (2005). *The software of the universe*. Aldershot: Ashgate.
Dorato, M. (2012). The natural ambiguity of the notion of "natural", and how to overcome it. *Epistemologia, 35*, 71–87.
Eisler, R. (1929). *Iesous basileus*. Heidelberg: Winter.
Eisler, R. (1946). *The royal art of astrology*. London: H. Joseph.
Franssen, M., Lokhorst, G.-J., & van de Poel, I. (2010) Philosophy of technology. In E. N. Zalta (Ed.), *The Stanford encyclopedia of philosophy* (Spring 2010 ed.). http://plato.stanford.edu/archives/spr2010/entries/technology/
Gigerenzer, G. (2008). *Gut feelings*. London: Penguin Books.
LeDoux, J. (1998). *The emotional brain: The mysterious underpinnings of emotional life*. New York: Simon & Schuster.
Longo, G. (2008), Impreparati dinanzi al "simbionte". http://www.swif.uniba.it/lei/rassegna/021119h.htm
Luria, A. (1969). *The mind of a mnemonist*. London: Cape.
Mach, E. (1896). *Populär-wissenschaftliche Vorlesungen*. Leipzig: Barth.

MacIntyre, A. (1984). *After virtue: A study in moral theory* (2nd ed.). Notre Dame: University of Notre Dame Press.
Mitcham, C. (1994). *Thinking through technology: The path between engineering and philosophy.* Chicago: University of Chicago Press.
Nozick, R. (1974). *Anarchy, state and utopia.* New York: Basic Books.
Pais, A. (1988). *Inward bound.* Oxford: Oxford University Press.
Putnam, H. (2004). *The collapse of the fact-value distinction.* Harvard: Harvard University Press.
Rodotà, S. (2004). Tra chip e sensori arriva il post-umano. *La Repubblica,* 6.12.2004.
Rossi, P. (1970). *Philosophy, technology and the arts in the early modern era.* New York: Harper and Row. (Eng. Trans. of *I philosophe e le machine.* 1962, Milano, Feltrinelli).
Sidgwick, H. (1907/1981). *The methods of ethics* (7th ed.). Indianapolis: Hackett Publishing Company.
Sober, E., & Wilson, D. S. (1998). *Onto others. The evolution and psychology of unselfish behaviour.* Cambridge, MA: Harvard University Press.
Thagard, P. (2011). *The brain and the meaning of life.* Princeton, NJ: Princeton University Press.

Chapter 12
Human Well-Being, Nature and Technology. A Comment on Dorato

Ibo van de Poel

Abstract In my paper I react to Mauro Dorato's contribution to this edited volume. In particular, I argue that two images of nature are at play in popular arguments about how technology may threaten nature, i.e. nature-as-environment and nature-as-essence. I critically review both images and discuss the potential normative force of arguments based on them. I argue for adopting human well-being rather than "human flourishing", as proposed by Dorato, as normative framework for evaluating new technologies as it avoids an essentialist reading of human nature. Moreover such a normative framework should also include other moral values like justice and sustainability.

1 Introduction

In his contribution, Mauro Dorato touches on a whole array of issues in the ethics of technology and develops challenging and interesting ideas. In particular he examines how the adjective "natural" is used in ethical controversies and the (lack of) argumentative power of such uses; he pleas for "human flourishing" as a normative framework for evaluating new technological developments and applies it to examples of implantable microchips. I start with some preliminary observations and critical remarks. I will then argue that two notions of nature are at play in Dorato's contribution and, in fact, in many public discussions about nature, namely nature-as-environment and nature-as-essence. I will consider the potential normative force of arguments based on these notions, and I will argue for human well-being, rather than the more specific notion of "human flourishing", as central moral value. This value, however, needs to be supplemented by other moral values in moral evaluations of new technology.

I. van de Poel (✉)
Section Ethics and Philosophy of Technology, Department Values, Technology & Innovation,
Delft University of Technology, Delft, The Netherlands
e-mail: i.r.vandepoel@tudelft.nl

2 Preliminary Remarks

Dorato starts with what are, according to him, two central questions in the philosophy of technology. The first one "involves the question whether it is technology or pure science that is the driving force of our increased understanding of the natural world." He does not discuss this issue in detail but I want to point out a remarkable feature in how he phrases the issue, namely that he tacitly supposes that technology is a form of applied science and that the question he mentions need to be understood in terms of pure versus applied science. However, the idea that technology is applied science is problematic and controversial, at least in the philosophy of technology and engineering.

The second issue is "the relationship between technology and values in general." Here Dorato refers primarily to the "ethics of technology." It should be pointed out, nevertheless, that there are also important questions about the epistemic values that play a role in technology. It is often argued that whereas science is about finding, or at least approaching, the truth, technology is about making things that work and that change the world. This suggests that the set of epistemic values that guides technology may well be different from that in science, or at least that what count as acceptable trade-offs between (epistemic, or epistemic and non-epistemic) values in science and technology may be different.

Let us focus, however, on the ethics of technology. Dorato draws attention to what he calls "the controversial notion of "human flourishing"" and "our emotional attitudes toward the dyad nature/technology", which "has been neglected in the analytic philosophy of technology." It is indeed true that there hasn't been much attention in the analytic philosophy of technology to how conceptions of nature on a more emotional or intuitive level affect moral debate; although there has recently been attention to the role of emotions in ethical judgment of new technologies (Roeser 2010) and also to how well-being (which I take to be a more general concept of which "human flourishing" is a more particular conceptualization) is important to evaluate technologies and engineering design (Brey et al. 2012; Desmet et al. 2013).

In the next part of the paper, Dorato examines six paradigmatic examples of appeals to nature as a source of ethical norms. These examples are very illuminating and he concludes that they are all "devoid of argumentative power, except, importantly, when it refers to a morally characterized human nature, and therefore to the notion of human flourishing." He then examines three possible explanations for the human tendency to derive norms from nature. Regretfully, it remains a bit unclear to what extent these explanations also offer a possible justification for deriving norms from nature. As he points out in his abstract, however, the most important is the first one that considers this tendency to be an adaptation.

Below, I want to look a bit more into Dorato's argument for employing "human flourishing" as a normative framework. It will become clear that the argument builds on a conception of nature-as-essence. I will criticize this foundation. First, however,

I want to look a bit deeper into the argument from adaptation, i.e. the argument that the human tendency to derive norms from nature is an evolutionary adaptation.

3 The Argument from Adaptation

Dorato summarizes what I would like to call the argument from adaptation as follows:

> Since keeping an equilibrium with our natural environment plausibly involves a certain invariance or stability of the niche in which we have lived for millennia, we probably evolved a universal attitude (which could have manifested itself in many cultures) to regard any radical change in our relationship with the environment as a potential threat to our survival. Since technology in particular nowadays is certainly perceived as a probable cause of such a change, technology is regarded as dangerous and threatening.

This is certainly a possible explanation for the tendency to derive norms from nature but could it somehow also be a justification? As Dorato alludes to, in some parts of environmental philosophy it may be considered a possible justification, but is it? To judge, I think we should first look a bit more carefully at the notion of nature that is at play here.

In his abstract, Dorato describes nature as "the complex of physical and biological regularities." This is certainly an important (descriptive) meaning of nature. However, it seems to me that in the explanation above a different notion or image of nature is at play.[1] This is the image of nature as environment. In this image, nature provides the environment in which humans live. Typically, this image presupposes a distinction between humans and nature: humans are not part of nature, rather they are a potential source of disruption of nature. At the same time, humans depend on nature-as-environment as a condition for survival. Humans therefore have reasons to see to it that their interventions into the environment do not undermine the capacity of that environment to sustain human survival.

A similar line of argumentation is visible in the argument from adaptation suggested by Dorato. I am inclined to believe that this argument has normative power *if* it is formulated in terms of humans and their environment. Humans obviously influence the environment in which they live and, obviously, this environment creates the living conditions for humans. The purpose of human survival thus gives us reasons to see to it that our interventions into our environment meet certain norms. However, it seems to me that the argument becomes much more dubious if it is coined in terms of technology and nature, at least if this is done according to the popular image of how technology may threaten nature-as-environment (Fig. 12.1). In this image, nature is seen as the environment, and technology is portrayed as part of human activities that (potentially) threaten nature.

---

[1] I use the term 'image' to describe a particular way laypeople or members of the public perceive nature. This does not necessarily amount to a neat and coherent conceptualization of nature.

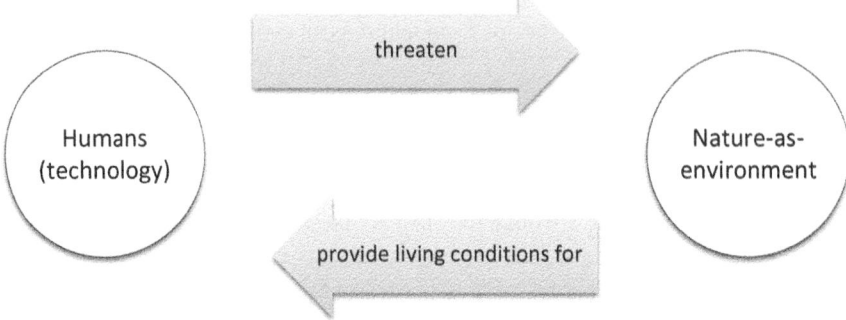

Fig. 12.1 A popular image of how technology may threaten nature-as-environment

This image of the relation between technology and nature seems to me problematic for three reasons. First, the separation of humans from nature seems problematic. Although we can, at least analytically, distinguish humans from their environment, such a distinction between humans and nature is problematic; humans are part of nature. Second, technology in this picture is solely placed in the human realm, while our daily experience suggests that technology has very much become part of our environment as well. Thirdly, although it is true that technological developments can threaten our environment, they can also be a source of sustenance or even improvement of the environment.

The human tendency to derive norms from nature may then, at least partly, be explained by our (far) past in which the *natural* environment was still the most important part of the environment humans relied on for survival. However, today, the human environment is to a large extent social and technological in nature. Of course, we still have reasons to adopt certain norms to ensure that our interventions in that environment at least sustain human survival (and the survival of other species). However, it is far from obvious that norms derived from nature are still instrumental in establishing such a relation with the environment given the substantial changes in our social and technological environment during the last two centuries.

4 Human Flourishing

Let me now turn to Dorato's arguments for human flourishing as central normative notion. A first thing to note is that the image of nature that is at stake here is rather different from that used in the argument from adaptation. It is based on an image of nature-as-essence rather than as environment. Dorato indeed speaks about "the realization of *our (alleged) moral essence*" (his emphasis) and even about "a natural tendency toward flourishing, unless a bad education distorts our "nature"."

Figure 12.2 tries to sketch the underlying image of nature-as-essence and how it may be threatened by technology as it seems to pop up regularly in popular

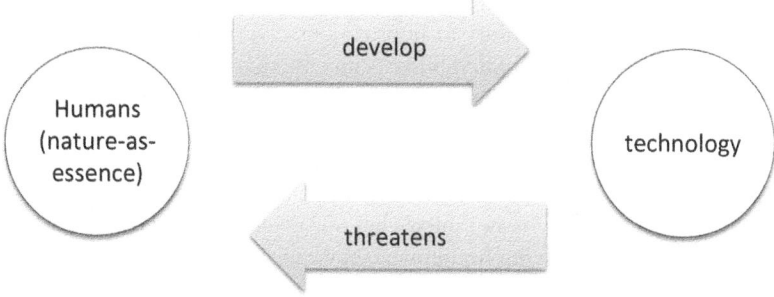

Fig. 12.2 A popular image of how technology may threaten nature-as-essence

discussions about technology. If we compare it to Fig. 12.1, it is interesting to see how technology and nature have switched places. Although nature-as-essence does not literally have a place, it is usually imagined (deep) inside humans rather than in the environment. Conversely, technology is seen as a threatening factor that may alienate humans from their true nature or essence.

This popular image may be criticized, I think, for at least two reasons. The first is that technology may not only threaten human nature but may also support it or even further it. Dorato is obviously well aware of this, as becomes clear in his discussion of the implantable chip cases. For him the criterion is whether technology contributes to human flourishing; and although he is aware that technology may threaten human flourishing, it is an empirical question whether it does so or not in a specific case.

A second objection to the popular image sketched in Fig. 12.2, however, seems to affect Dorato as well. According to this objection, the image separates human nature and technology too much. It can well be argued, I think, that in as far as we can speak of human nature or human essence, the deployment of technology is part of that nature or essence. Humans are tool-making animals, and arguably technology has always been part of humanity. It is true that in philosophy, technology is often portrayed as instrumental, or as an inferior way of access to reality or the truth. Technology is also usually not mentioned as an activity or capacity that contributes to human flourishing. However, I think one needs not to embrace transhumanism to see that an image that portrays technology only as a threat, or only as means, to human nature, rather than as a part of it, is too constraining.

I must admit that my own doubts about "human flourishing" as normative framework go deeper than that. I am weary of talk of human essence, certainly if it is intended as a foundation for ethical judgment. Of course, my weariness is in itself hardly an argument, so let me try to provide a philosophical argument. It seems to me that in moral debates, a reference to human (moral) essence is often used to conceal normative arguments and to stop debate. So, in the debate about the desirability of human enhancement through for example nanotechnologies, transhumanists may argue that it is part of human nature to develop technology and to continuously try to improve ourselves; if we take our moral nature seriously, we might even have a moral duty to improve ourselves. Opponents in the debate may argue that for example vulnerability is an essential part of human nature and that human enhancement threatens to destroy that quality; human enhancement according to such an opponent goes against human nature and should be avoided.

If this debate is conducted in terms of human nature or essence, it not only becomes irresolvable, but also the normative nature of the debate is concealed. What is at stake here in my view is not only what our given human nature is but rather what type of life is worth living for human beings, and how we should give shape to our lives so that they remain or become worth living. This is a normative issue that cannot be resolved by recourse to a given notion of human nature, but rather requires an articulation of what makes life worth living. I think such a debate could still be conducted in terms of "human flourishing" as long as that term is not naturalized to a description of human nature-as-essence. Since human flourishing is sometimes understood as a biological notion, I think it is better to use the more general term human well-being to avoid a naturalistic understanding.

For many practical ethical debates, my proposal to use human well-being as general normative framework to evaluate technologies is not that much different from Dorato's proposal to use "human flourishing" for that purpose. Both notions, well-being and flourishing, place emphasis on what makes one's life go well overall, and like Dorato I think that engagement in certain activities and the possession of certain capabilities are important here. The main difference is that where I would want to avoid a connection with talk about human nature-as-essence, Dorato explicitly makes this connection.

5 Human Well-Being

Let me now look a bit more precisely into the notion of human well-being and how it can be helpful in the moral evaluation of new technologies. It must be admitted that human well-being is itself a rather general notion that can be understood in different ways. In fact, in moral philosophy at least three main conceptualizations of human well-being can be found, i.e. (1) in terms of (pleasurable) experiences, (2) in terms of desire satisfaction and (3) in terms of so-called objective list accounts (Crisp 2013). The latter refers to a collection of values, activities or capacities that together contribute to, and constitute, human well-being.

Elsewhere I have argued for the adoption of an objective list account and have shown how such an account can inform the design of new technology (van de Poel 2012). Different authors have proposed somewhat different lists of items that constitute well-being. Griffin (1986, 67), for example, provides the following list:

1. Accomplishment
2. The components of human existence. This includes values such as autonomy, liberty and the basic capabilities to act

3. Understanding
4. Enjoyment (including perception of beauty)
5. Deep personal relationships

Nussbaum (2000) provides a list of capabilities that are constitutive of human well-being. What these accounts have in common is that elements on the list are not just means to human well-being, but rather constitutive elements of well-being. Dorato voices a similar idea:

> Cultivating genuine friendships and, devoting one's time to meaningful work, having a healthy parent–child relationship, or possessing literacy and education, are all objective goods for human beings, or part of what we mean by "flourishing", and are not just instrumental to it.

In fact, accounts of human well-being in terms of human flourishing often amount to objective list accounts that mention a range of virtues (or activities or capabilities) that are constitutive for flourishing.

But is human well-being the only value that is relevant in evaluating new technologies? I think that other values like safety, health, sustainability, justice and privacy are relevant as well. The first thing to note, here, is that human well-being itself is a composite value rather than one single value. With this I mean to say that human well-being exists in the realisation of a range of other values that together constitute human well-being. The list of values above mentioned by Griffin is a potential candidate for the values that are constitutive for human well-being.

I use the term 'constitutive value' here to distinguish it from so-called instrumental value. An instrumental value is a value that is valuable because it is a means to achieve another value. The relation between the instrumental value and the value that is achieved through it is causal. A constitutive value, however, has not just a causal but a conceptual relation to the value that it contributes to. If enjoyment is a constitutive value for human well-being, it is not just a means to human well-being (that could be replaced by other means) but rather a constitutive element of what human well-being means.

If human well-being is indeed a composite value that consists of a range of constitutive values, as I believe it to be, this has two important implications for the moral evaluation of new technologies on basis of human well-being. First, in many cases such evaluations will not take place in terms of human well-being *simpliciter* but rather in terms of the various constitutive values that make up human well-being (and that are affected by the technology at hand). Second, if technologies have different effects on the different constitutive values of human well-being, trade-offs between the various constitutive values can often not easily be made. The reason for this is that these constitutive values usually have intrinsic value due to their constitutive character for human well-being. Often a gain in one value, say accomplishment, cannot compensate for a loss in another value, say deep personal relations. Even if different constitutive values may all contribute to human well-being, there is no way that we can calculate the contribution of each single constitutive value so that a gain in one value evens out the loss in another.

I believe that in addition to the values mentioned by Griffin, such values as safety, health, and privacy are also constitutive values of human well-being that are important in the evaluation of new technologies. In addition to these values, there are values that are not constitutive of human well-being but are important in the evaluation of new technologies as well. One such a value is distributive justice. The value of distributive justice refers to a just distribution of certain goods among different people or different groups of people. The good can be the ability to fulfil one's needs, but also opportunity, income or human well-being. The different individuals or groups can live in one society, but also in different societies or in different generations (as in the case of intergenerational justice). Justice is an important value in addition to human well-being and not just as a contribution to human well-being.[2] The reason for that is that we have reason to strive for not just the largest amount of total well-being, but also for a just or fair distribution of well-being; we even have reason sometimes to accept less total well-being, if it leads to a more just or fair distribution of well-being. This suggests that well-being and justice are values that are independent of each other and cannot be reduced to each other.

There are, I think, other values besides justice and well-being that have a kind of independent status. One other candidate is sustainability. Sustainability can, at least partly, be understood in terms of (intergenerational) justice, but for some it would also refer to a kind of intrinsic value of nature. The notion of intrinsic value of nature is notoriously vague; if it is understood as stating that everything that is 'natural' is intrinsically good or valuable, it seems to me simply wrong, as Dorato's contribution makes very clear. However, I think one could still maintain that there is intrinsic value to biodiversity, and to the sustenance of certain species or ecosystems.

My aim here is not to develop a complete axiology of the values that are important in an ethics of technology. Rather I want to point out that a normative framework for evaluating new technologies should most likely contain more values than just human well-being (or human flourishing) and its constitutive values.

6 Implantable Microchips

The reader might wonder whether anything I have said would affect the moral evaluation of implantable microchips that Dorato offers. His evaluation is understandably quite sketchy. Still, it offers a good idea of the type of moral evaluation he aims at. One can see how in the different examples, different (constitutive) elements of human flourishing are at play, like the functioning of the human memory (and the importance of forgetting) and the nature of enjoyment. It seems to me that this way of proceeding underlines a point I made above: that we will usually not evaluate new

---

[2]Justice may also contribute to well-being because in a just society people may feel better or be more happy.

technologies in terms of well-being (or human flourishing) *simpliciter* but rather we will focus on the constitutive elements or values of well-being that are at play in the specific case.

If there is anything that I would criticise in Dorato's discussions of the examples, it would be the neglect of other values like, for example, privacy and justice that are relevant in evaluating possible applications and developments. Implantable microchips, and more generally possibilities for tagging people and things with Radio Frequency Identification (RFID) chips may give raise to serious privacy issues (Van den Hoven and Vermaas 2007). Applications of implantable microchips for medical or enhancement purposes raise not only questions about human-well-being but about justice as well, as such applications may affect the distribution of scarce and valuable capabilities like health and cognitive capabilities.

Acknowledgement Work on this contribution was supported by the Netherlands Organization for Scientific Research (NWO) as part of the research program 'New Technologies as Social Experiments' under project number 277-20-003.

References

Brey, P., Briggle, A., & Spence, E. (Eds.). (2012). *The good life in a technological culture*. New York: Routledge.
Crisp, R. (2013). Well-being. In E. N. Zalta (Ed.), *The Stanford encyclopedia of philosophy* (Summer 2013 ed.). http://plato.stanford.edu/archives/sum2013/entries/well-being/
Desmet, P. M. A., Pohlmeyer, A. E., & Forlizzi, J. (2013). Special issue editorial: Design for subjective well-being. *International Journal of Design, 7*, 1–3.
James, G. (1986). *Well-being: Its meaning, measurement, and moral importance*. Oxford: Clarendon Press.
Nussbaum, M. C. (2000). *Women and human development. The capabilities approach*. Cambridge: Cambridge University Press.
Roeser, S. (Ed.). (2010). *Emotions and risky technologies*. Dordrecht: Springer.
van de Poel, I. (2012). Can we design for well-being? In P. Brey, A. Briggle, & E. Spence (Eds.), *The good life in a technological age* (pp. 295–306). New York: Routledge.
Van den Hoven, J., & Vermaas, P. E. (2007). Nano-technology and privacy: On continuous surveillance outside the panopticon. *Journal of Medicine and Philosophy, 32*, 283–297.

Chapter 13
Philosophy of Science and Philosophy of Technology: One or Two Philosophies of One or Two Objects?

Maarten Franssen

Abstract During most of the twentieth century, philosophy of technology has been largely interested in the effects of technology on human society and culture whereas philosophy of science has focused on the content and justification of the ideas of scientists. During the past decades, a branch of philosophy of technology has developed that, similarly to the traditional orientation of philosophy of science, looks at technology itself and addresses the content and justification of the ideas and actions of engineers. In seeming opposition to the abundant evidence for a very intimate relation between science and technology, this 'internalist' philosophy of technology tends to emphasize how technology differs from science. In this essay I argue, against this contrastive characterization, that as practices science and technology can hardly be distinguished and that instances of the one rarely or ever occur without instances of the latter, and can hardly occur otherwise. This puts into question the rationale for maintaining philosophy of science and philosophy of technology as separate philosophical disciplines. Science and technology could perhaps be characterized ideal-typically as distinct activities, with science corresponding to the exercise of theoretical rationality and technology to the exercise of practical rationality. Even analysed in this ideal-typical way, however, science and technology can still be shown to share problems and partly overlap. It is questionable, therefore, whether the current separation of philosophy of science and philosophy of technology, grounded in conceptions of technology and science as contrastive, should be upheld.

M. Franssen (✉)
Section of Philosophy and Ethics, Faculty of Technology, Policy and Management, Delft University of Technology, P.O. Box 5015, 2600 GA Delft, The Netherlands
e-mail: m.p.m.franssen@tudelft.nl

1 Introduction

To any outside observer, science and technology must appear to be closely, even ever more closely related. So closely, in fact, that some people have proposed to speak of technoscience rather than the pair science and technology. No-one can fail to see that technology is deeply involved in the enterprise of science in that science's experimental practice is to a large extent technological. In July 2012 the discovery of the Higgs boson was first announced thanks to experiments performed with CERN's large hadron collider, which took 10 years to build and includes a tunnel with a length of 27 km dug at a depth range of 50–175 m below the surface, containing 1,600 superconducting magnets each weighing over 27 tonnes. With a costs of 7.5 billion euros this is probably the most expensive scientific instrument ever built.

It is equally hard to miss that science is part of technology's production of artefacts. No engineering curriculum exists which does not include courses in various branches of basic science, such as mechanics, thermodynamics, electrodynamics and quantum physics, from the very start. The term engineering science is used exactly for a form of science that exists as a constituent of (modern) technology.

Still, in apparent denial of these difficult-to-ignore observations, philosophically science and technology are sharply distinguished: they are supposed to have contrasting goals and correspondingly contrasting ways to achieve these goals. This antagonistic attitude does not come so much from philosophy of science, which still seems blithely unaware of a discipline called philosophy of technology,[1] but is wide-spread within philosophy of technology. Well-known characterizations from the early days are Skolimowski's claim that science concerns itself with what is and technology with what is to be, and Herbert Simon's claim that the scientist is concerned with how things are but the engineer with how things ought to be.[2] As recently as 2014, Galle and Kroes profess to this contrastive perspective by claiming that the subject matter of science is 'existing things', while the subject matter of design is 'novel things'. Not only does this contrastive picture seem to belie the actual state of affairs with respect to the relationship between science and technology, it also presents us with a conceptual problem, because the contrast class is identified differently in each of them: science is contrasted to technology, to engineering and to design. This apparent freedom in the characterization of the

---

[1] The 16-volume series of handbooks in the philosophy of science, general editors Gabbay, Thagard and Woods, published by Elsevier (North-Holland) between 2006 and 2009, includes, however, a volume on *Philosophy of technology and engineering sciences* (Meijers ed. 2009). This may indicate that the time of downright ignorance has come to an end. Developments within philosophy of technology during the past 20 years have contributed to this, one may assume.

[2] This is often how Simon is quoted, which may suggest that according to him technology is ultimately value-setting. Simon clarified the phrase, however, by adding "– ought to be, that is, in order to *attain goals* and to *function*" (1969, p. 5). Thus he clearly meant the invention and development of *means*, not the setting of ends.

technological side of things is part of what a philosophical comparison should address.

Given that about 50 years have passed since technology started to be characterized philosophically in this contrastive way with respect to science, and given that 50 years have been available for philosophical reflection on technology to develop this view, or to chart it and discover its limits – work which has accelerated during the past 20 years or so – this essay aims to question the way that philosophy of science and philosophy of technology have carved up the field of 'technoscience'[3] for philosophical reflection and to investigate whether perhaps another division of labour makes more philosophical sense.

2 Historical Differences Between the Philosophies of Science and Technology

One of the things that will strike anyone who travels in both disciplines is the difference in scope between philosophy of science and philosophy of technology. Philosophy of science is a reasonably focused affair. It deals roughly with just two issues, first that of how to articulate the method of science, its forms of reasoning, including the question whether there is such a method to articulate in the first place, and second the problem of scientific realism, the question to what extent the claims of science, in particular its theoretical claims, can be taken to be true, and whether any theoretical statement from science can ever be so taken. As Godfrey-Smith puts it, the goal of philosophy of science is to give a total picture of science that consists of two parts (2009, p. 102): "One part of the picture is an account of scientific practice, broadly understood. This is an account of how scientists develop and investigate ideas, which representational tools they employ, and how choices for one view over another are made. A second part, more philosophically tendentious, is an account of what all this activity achieves – how it relates to the world at large, what kind of knowledge it makes possible."

Note that this formulation does not mention truth, although the contrastive characterization of science and technology has also been made in terms of 'truth' being the primary goal or 'driver' of science, against 'usefulness' or 'success' or 'effectiveness' as the primary goal of technology (see e.g. Houkes 2009). This typology originates in philosophy of technology, not philosophy of science. It is a highly problematic way of presenting things, however, since the truth of most of the statements in which science deals, in particular of its quintessential statements – its laws and theories – is inaccessible. Even their 'approximate truth' is inaccessible, as the difficulties surrounding the notion of verisimilitude show. This forms an

---

[3]I stick to this term for want of a better one at the moment. The notion was coined by Bachelard and has found application in Science and Technology Studies, but not exactly in the sense I have in mind here.

important point of departure for philosophy of science, and should also inform us on how science and technology cannot be differentiated (more on this below).

Under the heading of philosophy of technology, in contrast, one can find discussed any range of topics that address the relation of technology to culture, society and, eventually, the 'essence of humankind', the relation of technology to science, and the relation of technology, so to speak, to itself. With the latter I refer to those aspects that most resemble philosophy of science: questions of methodology and of technology's principal goals or 'drivers'.

This difference in scope between the philosophical reflection on science and on technology arguably is a result of their quite different historical relation to philosophy. Philosophy of science, as well as science itself, continues questions that have been part of Western philosophy since its conception in Ancient Greece. The book that symbolizes the reaching of adulthood by science, Newton's *Mathematical principles of natural philosophy* (1687), likely received its title as a reply to Descartes' *Principles of philosophy* (1643). The differences between the two titles emphasize an important way in which science and philosophy would be different from that moment on, but at the time the two books were in the first place united by the co-occurrence of 'philosophy' in their titles. In contrast, philosophy and philosophers have never experienced similar historical ties to technology. From the moment the term Technikphilosophie and its later equivalents in other languages were coined, the topic attracted philosophers who were primarily interested in the way that technology shaped and changed the social and cultural life of humankind and the human condition,[4] without seeming to be much interested in its production and workings in relative isolation from its social dimension. These authors were, and their philosophical successors remain, interested in technological culture, technological society, technological man, technological morality, just not in technology itself, nor had any of them a first-hand experience with either engineering or science.

This particular conception of philosophy of technology is not included in what in this essay is referred to as philosophy of technology. Nevertheless it is a serious question whether the historical causes that I just presented for the different relations that science and technology have to philosophy also suffice to explain why a similar 'externalist' reflection on scientific culture, scientific society, scientific man and scientific morality was never included in the philosophy of science. That is not a question I will try to answer here, nor, for that matter, do I have any idea how to answer it.[5]

Apart from there being historical causes for the different orientations of philosophy of science and philosophy of technology, a comparison of science and

---

[4]This expression may owe its currency in philosophy to the 1958 book with that title by Hannah Arendt, one of the philosophers meant here. However, it had a predecessor in Lewis Mumford's 1944 book *The condition of man*. Note that the novel by Malraux, *La Condition humaine*, was translated into English twice, in 1934 in the US as *Man's fate* and also in 1934 in the UK as *Storm in Shanghai*, reissued as *Man's estate* in 1948.

[5]It could be argued that, for example, Ulrich Beck's *Risk society* belongs here.

technology and their philosophies has to take into account that the referents of the two terms are not invariably things of the same kind. There is a major ambiguity that both science and technology can refer to an activity or practice (to something process-like, therefore),[6] or to the results or products of that activity. Which is not to say that either term can refer just as often to an activity as to a product. Science, to my mind, primarily names an activity or practice; science is something you do. It is not immediately clear what the product of science as a practice is (more on this below), but the word 'science' is not often straightforwardly used to refer to it. At most it is used to refer to the sum total of what the activity of science, either as a whole or in a particular discipline, has delivered, the total body of scientific knowledge. In contrast, technology can clearly refer to a specific product of an activity or practice – in which case the noun is countable: we can speak of a technology and of technologies in the plural – or to the totality of these products in their social implementation (a crucial addition), in which case it is uncountable: technology as such.[7] Whereas science is not often used to refer to a product, technology is not often used to refer to a practice similar to or on a par with the practice of science. In that sense, as has already been noted, there are several competing terms, which both foremost indicate an activity or practice: design and engineering. Of these two, engineering is strictly a practice, like science, whereas design shares with technology the fundamental ambiguity between practice and outcome. We have design as designing, the practice, and we have a design or several designs, indicating products of that practice.

This leads to several questions. If we compare practices, then what is it that we compare science to: technology (if indeed 'technology' indicates a practice), engineering, or design? If we compare products, or compare goals with respect to output, do we compare technology or technologies to whatever science produces, or do we compare the products and the goals of design or of engineering? And what are these products?

However these questions are decided, I argue for the truth of the following statement: science and technology are not just two intimately connected practices or phenomena, but two almost completely interwoven practices or phenomena, to a point that it is hardly possible to have the one without the other and to be engaged in the one without being engaged in the other as well.

---

[6]The notion of 'activity' presupposes very little structure and is something that even an individual person could undertake, whereas the notion of 'practice' presupposes an engagement of many people and some form, however weak, of institutional organization, including ways in which the practice continues itself through the recruitment and education of new practitioners.

[7]In German and French, 'Technologie' and 'technologie' originally stood specifically for the systematic study of the phenomenon, either in its totality or in its separate manifestations as technologies; the phenomenon is referred to as 'Technik' and 'technique', resp. Likely due to the predominance of English, however, the words now tend to be used in the same way as in English.

3 Distinguishing vs. Separating Science and Technology; Theoretical and Practical Rationality

To see this, let us start by inquiring how relevant topics divide up over philosophy of science and philosophy of technology. To answer that, we can take as a starting point the following somewhat simplified version of the contrastive characterizations presented above, that science is about describing the world and technology about changing it. On the one hand, we cannot change the world without knowledge of what it is like: trivially we need to know the state from which we want to change it and we also need to have some minimal knowledge concerning the causal results of acting in order to make our 'changing the world' not something entirely random or arbitrary. The goal of changing the world has to be understood, to be sure, as changing it in a controlled, goal-directed way. Goal-directed action invariably requires knowledge as input, and the more detailed the specification of the goal, the more sophisticated and scientific the knowledge. It seems undeniable that in technology knowledge is applied, and that much of that knowledge is scientific knowledge. Which is not equivalent to the view, the straw man par excellence in this debate, that 'technology is applied science'. That view supposedly has technology import its knowledge from science, which is seen as an entirely separate discipline or practice. It ignores that insofar as scientific knowledge is applied in technology, this knowledge can be generated 'within' technology – where technology is viewed as an in some form institutionally delineated discipline.

On the other hand, finding out what the world is like turns out to be greatly facilitated by changing it locally in order to create a specific way for the world to be and observe that way. This is the experimental method, and setting up an experiment is by our general definition a form of technology.

That the two practices are interwoven does not preclude, however, that we can consider them apart, though this will have to proceed by surgical dissection, so to speak, as if separating a pair of conjoined twins. This inevitably results, however, in two highly abstracted, one might say ideal-typical characterizations of 'pure' science and 'pure' technology. It is even a fair question whether, in their pure form, these are the right labels.

In the case of 'pure' science the characterization is least controversial, in the sense of having a firm basis within philosophy: ideal-typically science is rational belief formation concerning the world. It falls under what philosophers call *theoretical rationality*, which contrasts with *practical rationality*, the rational formation of decisions to act or intentions to act.[8] Experimentation, that is, interfering with the world, enormously expands the extent of beliefs to be formed, but belief formation as such does not require it. Pure observation and reasoning can suffice, even for the production of scientific knowledge. The Darwinian theory of evolution

---

[8] In (Franssen 2006) I have suggested that the rational formation of desires, aims or goals could be seen as constituting a third independent variety of rationality, *orectical rationality*.

is based primarily on evidence gathered through unmediated observation. As soon as a theory is quantitatively articulated, measuring instruments, even of the most elementary form, have to be involved. Ptolemaic astronomy, for example, relied on instruments for the accurate measurement of angular separation of heavenly bodies. Still, measuring through instruments is not itself a way of changing the world or interfering with the world, though the introduction of a measurement instrument into the world obviously is.

We can then say that science is the systematic implementation of theoretical rationality, with practical rationality, in the form of technology or engineering, operative on the level of means, though theoretical rationality does not necessarily apply that means. In practice, however, theoretical rationality without experiments is as good as toothless. Accordingly the term 'science' is reserved for theoretical rationality that employs, as part of its rationality, technological means. It could be objected that not all experiments are technological in the sense that they involve engineered devices. For example, experiments on how crop yield depends on, say, sowing time need not do so. Still, as has already been suggested and as will be further discussed below, we can say that even experiments like these involve technology insofar as they involve a form of 'pure' practical rationality, a rational way of organizing things. Indirectly, Darwin also relied on non-technological experimentation, since past experience with the breeding of animals played an important part in Darwin's argumentation.

To characterize science as the implementation of theoretical rationality with respect to the phenomena of the world may be considered too broad, however. It would include police detective work, to establish who killed some murder victim, in science, which is not how we typically see it. This is not because this form of theoretical rationality is differently related to other forms: police work also relies on technology as a means: on ballistics, forensic medicine, fingerprint identification, and so forth. What makes science different from these instances of theoretical rationality is its aim of generality or universality. The conclusion that the police arrive at in a murder case – say, that John did it – has no validity beyond the particular case at hand; nothing follows from it with respect to any other aspect of the universe. Scientific conclusions, in contrast, are generally meant to have a significance beyond the particular circumstances in which they were (first) investigated; often even a significance throughout the universe: an explanation of how we come to see a rainbow in particular circumstances is valid whenever and wherever these or similar circumstances can be made to occur. However, to define science, therefore, as theoretical rationality aimed at the establishment of a particular type of beliefs – universal, general, law-like – runs into difficulties once we take the entire spectrum of the sciences in proper view. It ignores the historical sciences, which are much closer to police detective work than to physics. Even if we take into account that the English term 'science' has a much narrower field of application than the corresponding term in most other (European) languages – e.g. French 'science', German 'Wissenschaft'– and excludes the 'sciences of man' – then still we face the problem that evolutionary biology is to a large extent 'natural history', the theory of a single event, or chain of events, the origin and development of life on earth. Exactly to what extent science should be seen as narrower than theoretical rationality in the broadest sense, if at all, is an important topic that merits more attention.

A similar problem will emerge when characterizing ideal-typical technology, and it may be a problem that we cannot waive so easily there. Is it bringing into existence of artefacts? Or the solving of practical problems? Both have been proposed: the former by Galle and Kroes (2014), the latter by Houkes and Vermaas (2010). This will occupy us later in more detail. Let us for now characterize technology or engineering as the implementation of practical rationality aided by scientific means, although practical rationality can operate without scientific means, and rules for practical rationality are independent of the availability of any scientific means. Again, in practice, practical rationality without scientific (as it is understood here) support is as good as toothless, and accordingly the term 'engineering' is reserved for practical rationality that employs, as part of its rationality, scientific means.

Summing up, in either case, in the means, if they are scientific in support of technological ends, the technological is again present since these means are typically supported by technological 'submeans', and if they are technological in support of scientific ends, the scientific is again present because these means are typically supported by scientific 'submeans'. This brings out squarely how the two are interwoven at as many levels we care to distinguish.

This is also how these practices operate, that is, there may be a clear division of labour overall, but this division of labour is hardly stable at the level of individual people or teams. One person can switch roles easily, almost imperceptibly, in the course of research. A nice example is Newton's article in the *Transactions of the Royal Society* of March 1673, in which he describes his experiments with prisms and presents his theory of the mixed nature of white light, which consists of a multiplicity of rays differing in refrangibility. Once that conclusion has been presented, Newton immediately continues by pointing that this theory explains that there are fundamental limits to the resolution of telescopes and microscopes that use glass lenses and that this difficulty can be overcome by designing telescopes and microscopes such that they do not require light to be refracted. Thus was born the reflecting telescope. The switching from a descriptive and explanatory goal to a creative and innovative goal seems not to require any effort at all. A quite different case which nevertheless brings out the same point is the discovery of the splitting of the uranium atom as a consequence of bombarding it with neutrons, by Otto Hahn and Lise Meitner in 1939. Richard Rhodes, in his history of the development of the atomic bomb (1986), describes vividly how almost every physicists who learned of this result immediately recognized how this opened up the road to explosive devices with an unprecedented force, and just as immediately started to worry that this road would actually be travelled soon. (And history, of course, showed them right.) These examples do not put into question the distinction between theoretical and practical rationality, between describing the world and changing it. They do, however, show that these two are conceived, by people engaged in either of them, to be as intimately connected as can be, such that neither can exist without minimally tentatively bringing in the other.

We may have to conclude, therefore, that theoretical rationality and practical rationality are not so much practices themselves but analytical principles by which we carve up human practices – more restrictedly, as far as this essay is concerned, the 'total' practice of technoscience. Within that practice we can analytically prepare (in a sense analogous to anatomical preparation) 'subpractices', which are distinguished by whether at the highest level they are viewed as having a theoretical end – the formation of some belief – or a practical end – the decision for some course of action – but which at lower levels will involve applications of theoretical rationality and practical rationality, hierarchically ordered. What is more, the fact that a subpractice has been 'prepared' with either a theoretical or a practical goal at the highest level does not preclude that when the subpractice is viewed as it operates within the overall practice, the theoretical goal is put to the support of a higher practical goal, or the practical goal to the support of a higher theoretical goal. Take police detective work: we may suppose finding out who committed a crime, and how it was committed, to be the ultimate goal for the actual detective work, and the many detective series produced for television may confirm this picture, but of course settling the whodunnit question typically serves a practice of prosecution and punishment. Although it need not do so: finding out who killed John F. Kennedy, or how he was killed, or establishing the identity of Jack the Ripper, likely no longer will lead to prosecution, but the investigative work will not be any different for that. And further on a particular practice of prosecution and punishment can be used, and insofar as standards of rationality are still at work at that level will be used, to answer the theoretical question as to how particular such practices are correlated to societal crime levels and their development over time.

To insist, to return to matters squarely belonging to science and technology, that Hahn and Meitner's work on the fission of uranium atoms was science because their goal was a purely theoretical one, a search for knowledge and nothing but knowledge, would easily miss that such personal motivations do not clarify much concerning the nature of the activities engaged in. And, as we have seen in the case of Newton, such motivations can be extended at will and as it were instantaneously, even within a single individual, resulting in a practical motivation being put on top of a theoretical one, or the other way round.

In the following three sections, the question is taken up whether, and to what extent, philosophy of science and philosophy of technology can be meaningfully separated as the philosophy of theoretical rationality and the philosophy of practical rationality. In the first of these I argue that there are a number of philosophical problems that the two disciplines have in common, which puts into question the prospects of a neat separation of the two. In the subsequent two sections I argue that conceiving of philosophy of science as the philosophy of theoretical rationality and of philosophy of technology as the philosophy of practical rationality, respectively, will involve, if it is to succeed, major reorientations of the two disciplines as they are currently practised.

4 Issues Spanning the Philosophies of Science and Technology

Current philosophy of science is almost entirely dedicated to 'pure' science, that is, science considered without bringing into play its technical dimension and reflecting on that dimension as technology and therefore 'falling under' philosophy of technology. Philosophy of science is entirely concerned with how scientists arrive at their beliefs, in relation to how scientists ought to arrive there, and what the epistemic status of these beliefs are. This remains true even though during the past 25 years there has been a lot of interest into the experimental side of science. This interest, known as the 'new experimentalism', to which belong Mayo, Franklin, and Galison, to mention only a few, is not a reconsideration of the major issues of philosophy of science, but a reconsideration of how to address these issues, namely through looking more closely at experimentation. Lenk (2007, p. 87) refers to it as 'technologistic philosophy of science'.

Deborah Mayo, an important representative of the new experimentalism, described it as potentially offering a path between two contrasting positions that successively dominated the philosophy of science in the preceding period: logical empiricism (which treated the observational part of science as unproblematic) and post-Kuhnianism (which despaired of ever being able to vindicate science as a, or the, rational approach to the acquiring of knowledge about the world). The idea was to trade in the highly idealized notion of confirmation theory by the "actual procedures for arriving at experimental data and experimental knowledge". The key is Hacking's suggestion that "experiment may have a life of its own", where 'of its own' is interpreted as 'independent of theory'. Its agenda can thus be understood entirely as formed in response to discussions that were internal to philosophy of science throughout the second half of the twentieth century, and although not properly a continuation of these discussions, as occupying a terrain that formed as an effect of their whirlwinds. Mayo mentions as tenets of the new experimentalism (1994, pp. 270–271): (1) The aim of experimentation is not theory (dis)confirmation; (2) Experimental data can be justified independent of theory (or at least some theory); (3) Experimental knowledge is robust in the light of theory change. These can be understood a salvaging what could be salvaged from the general scepticism to which the post-Kuhnian positions gave rise. This work operates one level away from high theory, while still applying (non-Bayesian) probabilistic reasoning, for example to distinguish experimental artefacts from real signals, in contrast to the (Bayesian) probabilistic reasoning previously applied in confirmation theory.

In contrast, work in the philosophy of technology addresses technology both as 'pure' practical rationality, without taking into consideration the character of the knowledge that informs action, and as the specific form of practical rationality that is 'driven' just as much by its scientific knowledge base as by its ends. It even addresses that scientific driver itself – engineering science – independently of the actions to which it contributes, that is, as science. This is very different from the situation in the philosophy of science, where the new experimentalism does not address the technological drivers of modern science as technology, independently of the low-level empirical regularities and phenomena or the high-level models and theories that are candidates for confirmation.

Nevertheless, there are developments in philosophy of science that question the current division of labour between philosophy of science and philosophy of technology. This applies first of all to the current interest in the role of models in science. The starting point of this development is the work of Nancy Cartwright (1983) and Ian Hacking (1983), with a second stage initiated by the work of Margaret Morrison and Mary Morgan (1999). The common point is these views is dissatisfaction with the view, considered standard, that models in science are descriptive of the world, that they represent. Instead, argues Cartwright, models are carefully constructed toy worlds which capture a sort of scientific law-like order that is exactly absent in the real world. Morrison and Morgan emphasize the instrumentality of models; as they express it: "models have certain features which enable us to treat them as a technology"; they function as 'instruments of investigation' (1999, pp. 35 and 10 resp.). A similar view is expressed by Bas van Fraassen (2008, p. 238): "[T]heories are artifacts, constructed to aid us in planning and understanding [...]". This position adds a new dimension to the presence of technology in science: not only is technology materially present in the form of the engineered devices in experimental set-ups which feed the activity of theoretical rationality with data, though they are not actually a part of that activity, but through the use of models that activity itself becomes partly technological. It can be analysed as an activity in which instruments are used for certain ends. This line of reasoning is continued by Boon and Knuuttila (2009), who describe models as epistemic instruments. They do so particularly in relation to the use of models in engineering science, but their view does not seem to be restricted to this. There is an intuition that modelling plays a more predominant role in engineering than in 'pure' science, but whether that is so and whether that teaches us anything about technology or design in contrast to science is not clear.[9]

The view, however, leads to numerous questions. How exactly are we to understand the notion of 'instrument' and 'use' in the case of non-tangible entities, which models seem to be? Boon and Knuuttila solve this question by claiming that models are concrete objects (2009, p. 695) that are not only constructed but also manipulated. They subsequently specify the concreteness of models to mean that they are "concrete in the sense that they have a tangible dimension that can be worked on. ... when working with them we typically construct and manipulate external representational means such as diagrams or equations" (pp. 700–701). Apart from the question whether this explication retains the concreteness of models proper (to be taken up again below), this, however, may not be sufficient. It is questionable whether equations are any more tangible than models themselves. The

---

[9]Note that Boon has repeatedly (2006, 2011) argued for the view that engineering science is, indeed, science.

notion of 'mental use' here merits further analysis. It is unclear how that notion has to be understood in relation to models as instruments. Typically, an instrument is a device that a user applies to an object in order to have that object undergo a transformation. When I use a hammer to hammer a nail into a wall, I apply the hammer to the complex consisting of wall and nail, which I transform from being unconnected to being firmly connected.[10] None of the writings in which it is argued that models are instruments further analyse the use relation implied by this or discuss in any way the other relata involved in it. There is a tradition in model-based reasoning, but the instrumentalist view on models seems not to be connected to it. Only Giere occasionally refers to work in cognitive psychology, but with mixed appreciation; cf. his (1988, 1994).

Although the discussion on whether models can indeed be seen as instruments in science, including engineering science, addresses a methodological point, it was seen to lead to more fundamental philosophical questions concerning their ontological status. Through such questions the topic is connected to the most recent debates in the philosophy of science focusing on models, which concern ontology rather than methodology. The central question there is what sort of thing a model *is*. Unlike the methodological debate, which is restricted to science, including engineering science, the ontological debate has a counterpart in the philosophy of technology: the ontological status of *designs*.

In the philosophy of science, the past decade has seen a convergence to the view that models are not concrete objects, without, however, any form of consensus on what sort of entities they are if they are not concrete objects. Some defend the view that models are abstract objects (Giere 1988), some that they are fictional objects (Frigg 2010) or assets in a game of make-belief (Toon 2010).[11] It was already discussed that the position that models are non-material entities makes it problematic to uphold the idea that models are instruments, that we *use* models. For models in science, this is part of a larger discussion. The contrastive position is that models refer, to their target systems, or anyway to (aspects of) the world. Traditionally, that is our conception of language: words, at least many words, and sentences, at least many sentences, refer: to objects or properties or events or states of affairs. Can we not say that we use language to refer, and that therefore we use abstract entities to refer? It is not clear, however, that language deals in abstract entities. Spoken words and sentences certainly are not abstract, nor are samples of written or printed text. It is the types that are abstract; but this relates to well-known discussions of universals and (natural) kinds, not to be taken up here.

---

[10]Colloquially, just the nail will be singled out as what the hammer is applied to – brought down upon – but this ignores that the desired effect will only occur when this is done while the nail is pressed orthogonally or obliquely, but anyway firmly, to the wall.

[11]Note that the status of fictional objects, and the related but larger category of intentional objects, is still a topic of wide controversy in metaphysics, although the claims defended with respect to models in science and technology are much informed by work done in metaphysics; see Thomasson (1996) and Kriegel (2008) for two recent presentations of opposing positions.

The related discussion in philosophy of technology is the status of designs, that is, the conceptions of artefacts emerging from the activity of designing. That activity need not necessarily lead to a realized artefact in order to count as design; it should, however, lead to a 'plan' of an artefact. What exactly that is remains unclear, except that it is something non-material. Kroes (2009), for example, states that the activity of design need not end in the actual realization of a copy of the designed artefact; the 'true' result of design as an activity, that is, a design as a product, is a mental object. How to conceive of this mental object, however, is problematic; it is described as a plan for an artefact and as a representation of the artefact. As a plan, however, it is not necessarily a plan for making the artefact, though that could be included. This leaves us with two problems: what a plan of an artefact is if not a plan for making it, and how a plan can represent.

The characterization of designs as plans in a way addresses what is at issue in this chapter: the commonalties of and differences between philosophy of science and philosophy of technology. By calling a design a plan, there is the suggestion that we are in the realm of practical rationality. No one has proposed to conceive of models as plans, not even in the context where models are conceived as instruments. By claiming that the plan represents an actual artefact, or rather an artefact kind, we are in the realm of theoretical rationality, where representation is the paramount relation supposed to exist between models and their targets. To conceive of this relation as representation, however, is highly problematic. Models that are false, in containing non-existent entities (e.g. the ether, phlogiston, absolute space) cannot properly be said to represent anything. Since whether or not this is the case typically becomes clear only long after these models have served their purpose, an account of how models figure in science cannot depend on seeing them as representing. At most, they can be seen as being aimed to represent.

The situation in design is similar. Since a design precedes any actual manufactured artefact, there is no existing object to be represented by a design. This is a general point, extending to views that present designs as *descriptions* of artefacts (e.g. Houkes and Vermaas 2010). How description is related to or differs from representation is anyway an open question with respect to these views, but however we see this, the connection between the two characterizations of design, one a plan, the other a representation, is problematic as well: how does a plan represent? It seems to be the wrong kind of thing for representation. The divide between theoretical and practical rationality cannot be straddled so casually. The closing remark of Kroes's discussion of design, that "no clear analysis of the notion of design of a technical artefact has yet been provided" therefore seems entirely accurate.

5 Isolating Science as Theoretical Rationality, and Its Problems

It has been argued that the nature of the knowledge sought, and the process for acquiring that knowledge, are different when in the service of a practical goal. Examples are the method of systematic parameter variation in order to establish where some maximum or minimum of a critical behaviour occurs. Such methods are said to be used particularly in circumstances where no theory is available, in order to get on with a design job (so Hendricks et al. 2000). Vincenti discusses such methods in association with 'what engineers know'. However, the knowledge that this results in is still knowledge, that is, can be phrased in terms of belief states – e.g. 'lift on a rectangular surface with constant horizontal velocity between x and y is maximal for shape z_i among shapes z_1 to z_n'. A discussion of the extent to which such emerging beliefs are justified, how to avoid accepting experimental artefacts, and so forth, is continuous with what is discussed within philosophy of science, e.g. in relation to the 'new experimentalism'. Once we recognize that in the practice of doing science there is nothing that can discriminate between being interested in the *truth* of a belief and being interested in its empirical adequacy, its 'holding' within a range of circumstances, no further criteria are available for discriminating between scientific knowledge and technical knowledge, as long as we are interested in knowledge claims, that is, descriptive claims. Beliefs about how the world reveals itself at a certain level of detail, within a certain range of variation for particular parameters, are still beliefs about what the world is like. To be sure, the nature of practical problems will often go together with a particular type of knowledge, where 'type' refers to scope, detail, and so forth. This does not warrant a 'scaling up' of these type differences to one type being and the other type not being scientific knowledge, that is, subject to the standards of justification and the patterns of reasoning that are studied in philosophy of science, or even descriptive knowledge per se, that is beliefs that, when articulated with care, are candidates for being true or false.

Seely (1984) describes an amusing case where two parallel investigations were undertaken in the USA in the interbellum to increase road quality. The federal government's Bureau of Public Roads initiated a research project that was based on the hypothesis that the main causes of road damage are vertical forces, caused by cars and lorries being lifted slightly by irregularities in the road surfaces and than coming down again. The investigators then aimed to establish empirical laws containing the relation between the size and frequency of these forces and the damage caused in slabs of particular materials used for road surfacing, In order to find these results, they built a huge machine that would bump such slabs continuously, controlling for size and frequency, after which the damage to the slabs would be determined. The experiment dragged on for considerable time due to the difficulties in making the experimental set-up work, that is, achieve a sufficient constancy of force and frequency for a sufficient length of time, and to the difficulties in selecting criteria for the damage done. In the meantime the Illinois State Highway Department, becoming impatient, set up an experiment of its own; it built a track of several kilometres, made up of stretches of various materials available for road surfacing, and then had a caravan of lorries circuit this track for months in a row. This sufficed for being able to pick the best candidate to be used for road surfacing. This case makes clear that an important aspect of practical rationality is to recognize the accuracy of the knowledge required for solving a practical problem – choosing a course of action – and the most efficient way to acquire that knowledge. But it does not allow us to discriminate between the types of knowledge produced by the Illinois State Highway Department and (eventually) by the Bureau of Public Roads as being, in the former case, technical or practical knowledge and, in the latter case, scientific ('unpractical') knowledge. In both cases we are dealing with the acceptance of certain belief states as being adequately supported by empirical evidence; differences exist only in the match between the nature of the belief claim and the sort of support required for it. In both cases the researchers were driven by practical goals, and the Illinois researchers were not less engaged in doing (engineering) science than the federal researchers. Even the results of the State of Illinois' research has a substantial level of generality: it is assumed that the performance of the various surface materials when driven on remain valid at other times and places.[12]

It may be thought that the situation cannot be as symmetrical or neutral with respect to the theoretical-practical distinction as here suggested. Is the overall practice of technoscience not ultimately practically oriented? Is that not exactly what the term 'practice' implies? This is an important point, and one with a wider philosophical significance than may initially seem.

The basic opposition between describing the world and changing is one of overall goals. Setting oneself goals does not, of course, guarantee that these goals are attainable. The goal of describing the world has traditionally been interpreted as requiring that science produces claims about the world that are true. Accordingly, truth is often presented as the goal of science, or its main 'driver'. It has long been recognized however, that insofar as the descriptive claims of science go beyond stating merely empirical phenomena or states of affairs, stating laws and theories – systems of interconnected laws –, their truth is inaccessible to us.[13] Instead, therefore, science must aim for something else, which is typically phrased as empirical adequacy, a match between what we observe the empirical facts to be and what the laws and theories imply with respect to these facts. Philosophy of science is more or less defined by the lack of consensus that exists concerning how this match has to be 'measured', and how good it must be, or what the status is of the laws and theories that pass whatever test is used. Against the philosophers who would want to claim that a theory's empirical adequacy is a sign of its truth, or

---

[12] There is a caveat concerning seasonal influences, depending on whether the testing spread over an entire year.

[13] Typically Hume is credited as the first philosopher to hammer this message home, but actually the insight dates back to antiquity, when it was known as skepticism. The response, therefore, that science must settle and can settle for something less than truth has a pedigree that goes back much further than many people are aware of, having been defended in the seventeenth century by Gassendi; see e.g. Fisher (2005). A related question is whether even for singular statements, which report phenomena or states of affairs, truth can be claimed. This question has been vividly discussed during the past one and a half centuries. The early radical positions, where truth receded to include in the end just subjective sensory experience, resulting in truth becoming impotent in the process, are no longer popular, but the issue continues to be played out against empiricist positions like, for example, Van Fraassen's.

indicates its likely truth, others hold that the concept of truth has no place in science beyond the truth of the statements that describe the observational facts. This latter position is often called 'instrumentalism'.

This must make us aware of the fact that the general characterization of science proposed until now, as being engaged in describing the world, may not be general enough; a more general characterization would be that science 'accounts for' the world. Associated with this is a similar unclarity concerning what the products of science are, i.e. in what consists this accounting for the world. Those who hold that the goal of science is literally to describe the world must hold that these descriptions are what science must deliver, and that the laws and theories that science 'uncovers' exactly qualify as these descriptions.[14] For those who subscribe to this position our inability to arrive at the truth of these laws and theories must be a major problem, if not the major problem, in philosophy of science. Instrumentalists, in contrast, hold that the goal of science is to 'account for' the world in the form of allowing us to predict and explain singular facts. Laws and theories are mere instruments to arrive at accurate predictions; the question of their truth never enters the picture. Insofar as instrumentalism aims at both prediction and explanation, it requires a rethinking of the notion of explanation. The colloquial understanding of that notion will require, for example in the case where the phenomenon of the rainbow is explained by bringing in the laws of optics and the nature of white light as discovered by Newton, that this counts as an explanation only if the story, including its theoretical claims, is true. This recourse to can only be avoided by reinterpreting the notion of explanation as being 'nothing but' a variety of prediction, as was indeed part of the deductive-nomological account of explanation defended foremost by Hempel.

The instrumentalist conception of science, including the term 'instrumentalism' itself, may suggest that the enterprise of science is lifted in its entirety from theoretical rationality and placed within practical rationality. This was not the case for the classical form of instrumentalism, which was closely linked to logical empiricism of the first half of the twentieth century. The goal of prediction was held to be a cognitive goal, the acceptance or rejection of predictions were seen as governed by cognitive criteria characteristic of science and, most importantly, such acceptances were not conceived as *actions*. A reinterpretation of belief acceptance as being a variety of practical rationality has come to be argued recently, however, under the heading of 'epistemic instrumentalism'. Epistemic instrumentalists (see e.g. Stanford 2006) do look upon the acceptance or rejection of beliefs concerning the world as actions, and hold that how we should make choices with respect to what to believe should be part of the general framework of practical rationality, where courses of action are rational if they tend on the whole to satisfy the totality of our goals best. It is, moreover, characteristic of most current theories of practical rationality that these goals themselves fall outside the scope of practical rationality. Perhaps not entirely – can it be rational to aim for both p and not-p being the case? –

---

[14]Galle and Kroes seem to belong here; cf. their claim (2014, p. 221): "What a scientist must do in order to 'do science' is ... essentially to produce a scientific theory."

but to a sufficient extent so as to allow that different people, when confronted with the same empirical evidence, nevertheless come to accept different beliefs about what the world is like, simply because they differ in the aims they seek to realize.

Against epistemic instrumentalism Kelly (2003) argues that whether or not one has reasons to believe, say, p, has nothing to do with the goals one has, that is, one has these reasons categorically, in the Kantian sense of non-hypothetically. We could say that q is a reason for A to accept p without having any idea of A's goals, epistemic or not. He discusses as an example someone who does not want to know how a particular movie ends. One has good reason to avoid evidence for the end, but once the evidence is there, is it rational to reject the belief about the movie's end that it furnishes? The mechanisms of belief formation are such that it cannot entirely count as the implementation of practical rationality, i.e. choosing a course of action. It is minimally partly contained in the making up of one's mind that precedes choosing, something that, notwithstanding the term 'making', is not practical. If the belief requires effort, that is, performing an action that is a means to the end of having the belief, then obviously it is rational to reject going through the effort so as not to arrive at the unwanted belief. However, if the formation of the belief, once the evidence is in, is 'automatic' and the acceptance is a form of bookkeeping about where to file the belief, then it is irrational to refuse to file it. It would be creating a split between knowing that p and admitting that one knows that p. Surely we would judge a person irrational who, as we know, has been presented with knock-down evidence concerning the end of a movie and who continues to declare that she does not know the end because she saw no reason to draw the obvious conclusion from the evidence. And we would do so even in the case where that evidence has been carefully fabricated so as to present the end of the movie falsely. It seems precisely characteristic of theoretical rationality that there are epistemic norms that govern us regardless of our goals; we are subject to these norms merely as believers. One cannot rationally continue to believe not-p in the face of overwhelming evidence in support of p. Even if it later turns out that p is false after all, this does not make the premature acceptance of not-p more rational.[15]

This, however, applies to beliefs about states of affairs the truth of which is accessible. As stated earlier, the beliefs that are central to science are inaccessible as regards their truth. Empirical adequacy is the best we can do, and as is well known, theories are underdetermined by the empirical evidence. The acceptance or rejection of scientific beliefs has to proceed under other criteria than truth, therefore, and there are several. In current discussions, these are referred to as epistemic virtues – next to empirical adequacy they comprise for example simplicity and coherence with other accepted theories.

---

[15] Some disagree. For a discussion of this point see Franssen (2009). Note, moreover, that the difficulties met by this recent epistemic instrumentalism do not invalidate traditional instrumentalism as a viable position concerning what science is all about, next to realism, since it respects the constraints of theoretical rationality. During the past decades many views have been proposed that try to steer a midway course between these two extremes. This essay is not the place to discuss these proposals in detail.

This again brings the philosophies of science and technology closer together. Kroes (2009) emphasizes that the decision-making aspect of design is what distinguishes technology from science. Such decisions are to a large extent underdetermined by the problem definition; they involve trade-offs, decisions to redefine the problem by modifying the list of functional requirements. Decision making involves just as much the creation of options to choose from as the choice among the options. That is what makes engineering design invention rather than discovery. The latter would imply that there is one 'true' solution that is already implicitly there in the problem definition. The range of considerations for any design problem includes costs, safety, sustainability, aesthetics and social impact. There is no a priori fixation of where the limits are for taking something into consideration.

Given the underdetermination of theories by the empirical evidence, the acceptance or rejection of a theory, or of an explanation or prediction based on a theory or model, also has the character of a choice to some extent (as Kroes acknowledges). This means that, if one wished to retain the distinction between theoretical and practical rationality, the mere occurrence of decision making, in the form of a choice between alternatives, cannot be defining of practical rationality. It might be thought that 'theoretical choice' is characterized by choice criteria of a particular kind, the epistemic virtues mentioned above. If it may seem that criteria typical for practical rationality play a role as well, such as costs, safety and social impact, this concerns only the technology, in the form of experimentation, brought in to support the solution of the theoretical problem. Such considerations may have a causal influence on the outcome of the theoretical decision problem. To give just one example, in his analyses of the controversy over the charge of the electron between Robert Millikan and Felix Ehrenhaft during the first two decades of the twentieth century, Holton (1978) mentions that the different results that the two scientists obtained were partly due to the technical sophistication of the instruments they built and used, with Millikan's being by far the cheaper one, but by coincidence just of the right simplicity for what he had set out to find.

That does not make cost a criterion in theory choice or belief acceptance. It has been argued, however, that social impact is such a criterion, or at least should be. In 1953 Richard Rudner defended the view that exactly because hypotheses and theories, are seldom if ever proved true by the evidence, scientific knowledge, in the form of accepted and rejected hypotheses and theories, is fallible. Science may accept a hypothesis although it is in fact false, or it may reject a hypothesis although it is in fact true. But accepting false hypotheses generally has consequences – predictions that fail – and so does rejecting true hypotheses – failures to predict – which affect the well-being of people. Therefore in decisions to accept or reject hypotheses and theories, the wider consequences of having it wrong must be taken into account. Against Rudner's view, a view defended again more recently in greater detail by Douglas (2000), Jeffrey (1956) argued that the argument does not go through because science does not accept or reject hypotheses but only 'calculates' the amount of support that the available evidence grants to theoretical claims. This controversy, then, points out from another angle that there is fundamental unclarity about the products of science. To the ambiguity already noticed – whether they are laws and theories or predictions – now a further ambiguity is added – perhaps these products are rather degrees of belief in laws and theories, or alternatively likely candidates for predictions and explanations. This shows that the question how the analytical principles of theoretical rationality and practical rationality carve up the practice of technoscience does not have an easy answer.

6 Isolating Technology as Practical Rationality, and Its Problems

Let us finally reverse the issue to the products of technology, a question that has already been briefly addressed above. If we let ourselves be guided by the distinction between theoretical and practical rationality, then we are here facing the question what to do, rather than the question what to believe. The traditional core business of technology, the design and manufacture of artefacts, would match this, since it can be seen as being ultimately guided by the question what to deliver in response to a request for an artefact that satisfies certain functional requirements, a request that comes from outside of the practice of technology proper, from the clients and customers that make up society.[16] This leads right away to the following observation. First, it places the motive force of technical development elsewhere, outside of technology itself, quite different from science. This, however, may merely reflect a false self-image of technology, a failure to acknowledge that many if not most design problems find their origin in technology itself, and 'customer needs' are de facto 'perceived customer needs' or 'assumed customer needs'. Second, and more seriously, within the context of this essay, it makes technology represent only a small part of the implementation of practical rationality. I argued above that some restriction may also exist for science with respect to theoretical rationality, where science only concerns beliefs which contain an element of generality, but at the same time I questioned whether this can be upheld once we include in science the study of history. Here, however, we are facing a much more substantial restriction. Third, finally, and even more seriously, it makes it questionable whether technology does in fact belong to practical rationality, since it is not clear to what extent technology is engaged in doing things. The actual use of the artefacts designed within technology, the solution of practical problems, even the manufacture of artefacts by industry, need not be included within technology. Compare, for example, the view of Galle and Kroes (2014), already referred to above, that the activity of design need only result in a *proposal* for a design (whatever that is precisely). Such a product could easily be redescribed as a theoretical or cognitive claim, namely that a particular configuration is the best possible realization of a list of functional requirements for an artefact from among a set of candidates. Of course, as Kroes (2009) emphasizes,

---

[16]The standard picture of the phases of the design process, as it is taught in engineering methodology, starts with the phase where the 'customer needs' are articulated; see e.g. Suh (2001).

these candidates have to be invented. That, however, would merely be pointing out that practical rationality is at work in the means to arrive at a theoretical end: the claim as stated. We have also seen that there may not be a principled distinction between the sorts of constraints and criteria that govern the acceptance of beliefs in science and technology. In this way it would be very hard to maintain a categorical distinction between science and technology as implementations of theoretical and practical rationality, respectively.

The strong association of technology with artefacts – material devices – and through them with engineering is arguably untenable. Not mainly because it leaves so much of practical rationality unaccounted for – whether technology *is* the implementation of practical rationality is still a research question – but because it leaves much of the scope of design unaccounted for, and much of science-backed problem-solving. Take, for example, the practices of medicine and of economics, insofar as the latter aims to design private (business) and public (government) policies.[17] Economics depends hardly if at all on engineered artefacts, and medicine depends foremost on drugs, the artefactual status of which is problematic.[18] And even within institutionalized engineering the engineering artefact is becoming less central, giving way to a much larger class of 'solution concepts': what technology delivers comprises organizational schemes, management models, rules, procedures, recipes, systems.[19]

There are two recent contributions, from the philosophies of science and technology, respectively, that aim to steer the characterization of technology and design away from artefacts, although both, I argue, do so only half-heartedly.

In the philosophy of science Ilkka Niiniluoto (1993) has tried to clarify the distinction between science and technology by accepting that in science one is after truth whereas in technology one is after effectiveness. Technology designs artefacts (which to him can be material as well as social) to serve as tools in the interaction with and transformation of reality. This distinction, however, according to Niiniluoto, does not yet settle what logical form the knowledge must have which research must produce in order to arrive at what the practice is after. In the case

---

[17]In 2001 Willem Wagenaar, the dean of the University of Leiden in the Netherlands, and a world-renowned psychologist, argued in a lecture in Delft (home to the largest Dutch university of technology) that exactly for this reason the distinction between 'regular' universities and universities of technology, and ultimately between science and technology, is empty; see Wagenaar (2001). Why the application of medical and economic knowledge is not called engineering is a philosophical enigma, that is, none of the more obvious criteria – presence or absence of the application of theories from the natural sciences or the application of mathematics or modelling – suffices to explain this. To me this is nothing but an historical accident.

[18]They are artificial, in the sense of synthetic, as samples or even molecules, but they are not artefactual as kinds: they form natural kinds. See for more on this (Franssen and Kroes 2014).

[19]This ties in with Simon's view of the content of the 'science of design'; see his (1969), pp. 79–80. For an approach that puts systems central, not artefacts, see Franssen (2014). For a historical study that emphasizes that organization has been a crucial ingredient of technological innovation from the start, see Hughes (1983).

of science, there is consensus that this knowledge, the totality of the products of research that is aimed for truth, consists of laws and theories. Niiniluoto proposes that in the case of technology the knowledge, the totality of the products of research aimed for effective artefacts – design science or design research as he calls it – consists of technical norms. An example of a technical norm is 'In order to make the hut habitable, it has to be heated'. This is a prescriptive statement, which has to be sharply distinguished from the statement 'The hut can only be made habitable if it is heated'. That sort of statement, articulating a necessary causal connection, is an outcome of scientific research. Nevertheless Niiniluoto holds that technical norms have a truth value just as well.

By insisting that technical norms are candidates for being true or false, which allows him to classify them as knowledge, and by insisting that systematic 'science-like' research in support of design should result in knowledge, Niiniluoto may exactly be missing what is at issue here. It is significant that Georg Henrik von Wright, who introduced the notion of a technical norm in his *Norm and action* (1963), did not believe that technical norms have a truth value. It may be that the insistence that effective interference with the world proceeds through artefacts obscures from view the more fundamental aspects of interfering with the world and trying to do so effectively. This is not to deny that artefacts are in fact ubiquitous in this role; it is to deny that this ubiquity is what an analysis of practical rationality must focus on. Theoretical and practical rationality are at work in science and technology, and much what concerns the design of artefacts may in fact fall under theoretical rationality.

Similar problems crop up for a quite different proposal from the philosophy of technology. Houkes and Vermaas (2010) have been advocating, since about a decade, what they term an action-theoretic approach to technical artefacts and their functions. Central to their approach is the notion of a use plan. This notion, however, is not tied to the concept of an artefact. Plans are for the solution of practical problems: that is where the notion of a plan first emerges.[20] A plan may contain the manipulation of some tangible object, in which case it must contain as a 'subplan' a *use plan* for that object, and if the object is as of yet non-existent it must contain as another 'subplan' a *make plan* for that object – this is where traditional product engineering comes in. By terming their basic plan notion also a use plan, however, they partly invalidate their analysis. They fail to sufficiently dissociate, through their action-theoretic approach to design, technology and design from artefacts. Plans as solution concepts for practical problems do not necessarily involve the use of artefacts, or the use of objects at all. My claim is that the firm connection to artefacts is exactly what is holding up a systematic analysis of technology as an implementation of practical rationality, and an unwarranted identification of engineering, technology and design as basically the same activity or phenomenon.

---

[20]Cf. the title of (Houkes 2008): 'Designing is the constructing of use plans'. The choice of the term 'constructing' brings us back to the discussion on whether design and use is restricted to concrete entities.

There are many 'things' that we may 'use' – we may, for instance, create and use 'situations', as any illusionist or pickpocket will acknowledge. And even with this broad reading of 'using things' we have not exhausted the scope of 'doing things', 'bringing things about', 'making things happen'. Even the use plan for such a simple tool as a hammer will not suffice for hammering a nail in a wall: that use plan needs to be embedded in a larger plan, which will specify where to procure the nail, what nail to choose, and which spot on the wall to choose for hammering the nail in. Referring to this larger plan also as a use plan, where the notion of use plan is specifically introduced as the tool through which artefacts and artefact functions are to be understood, rather than as a tool through which practical problems and solutions to such problems are to be understood, obscures rather than clarifies.

Even though their action-theoretic approach may be exactly what we need, given that we are trying to come to grasp with practical rationality, much work still remains to be done. Also the notion of 'a design' remains floating. Is any design a plan? If so, is any plan a design? The notion of 'plan' may still need further analysis – Houkes and Vermaas seem to rely entirely on Bratman's work – and may need to be connected to the more extensive philosophical literature on rules.

7 Conclusion

This essay presents an overview of ways in which the current literature in the philosophy of science and the philosophy of technology captures aspects of the practices of science and technology and is able to account for their features and the extent to which these features capture essential aspects of these practices and their interrelation. This overview teaches us, I argue, that not only are these practices extremely interwoven in their actual manifestations, but the attempts to analyse them even ideal-typically as implementations or manifestations of theoretical and practical rational have revealed numerous interconnections and common features. So many, in fact, that we are perhaps better advised to trade in the two philosophical subdisciplines, operating at a vast distance from each other, for a single philosophy of technoscience.

Acknowledgement I thank Peter Kroes, Sjoerd Zwart and Sven Ove Hansson for their helpful comments on a draft version of this chapter.

References

Boon, M. (2006). How science is applied in technology. *International Studies in the Philosophy of Science, 20*, 27–47.
Boon, M. (2011). In defense of engineering sciences: On the epistemological relations between science and technology. *Technè, 15*, 49–71.

Boon, M., & Knuuttila, T. (2009). Models as epistemic tools in engineering sciences. In A. Meijers (Ed.), *Philosophy of technology and engineering sciences* (Handbook of the philosophy of science, Vol. 9, pp. 693–726). Amsterdam: North-Holland.

Cartwright, N. (1983). *How the laws of physics lie*. Oxford/New York: Oxford University Press.

Douglas, H. (2000). Inductive risk and values in science. *Philosophy of Science, 67*, 559–579.

Fisher, S. (2005). Gassendi. In *Stanford encyclopedia of philosophy*. http://plato.stanford.edu/entries/gassendi/. Text as revised 2013, first published 2005.

Franssen, M. (2006). The normativity of artefacts. *Studies in History and Philosophy of Science, 37*, 42–57.

Franssen, M. (2009). Artefacts and normativity. In A. Meijers (Ed.), *Philosophy of technology and engineering sciences* (Handbook of the philosophy of science, Vol. 9, pp. 923–952). Amsterdam: North-Holland.

Franssen, M. (2014). Modelling systems in technology as instrumental systems. In L. Magnani (Ed.), *Model-based reasoning in science and technology: Theoretical and cognitive issues* (pp. 543–562). Heidelberg etc.: Springer.

Franssen, M., & Kroes, P. (2014). Artefact kinds, ontological criteria and forms of mind-dependence. In M. Franssen, P. Kroes, T. A. C. Reydon, & P. E. Vermaas (Eds.), *Artefact kinds: Ontology and the human-made world* (Synthese Library, Vol. 365, pp. 63–83). Cham etc.: Springer.

Frigg, R. (2010). Models and fiction. *Synthese, 172*, 251–268.

Galle, P., & Kroes, P. (2014). Science and design: Identical twins? *Design Studies, 35*, 201–231.

Giere, R. N. (1988). *Explaining science: A cognitive approach*. Chicago: University of Chicago Press.

Giere, R. N. (1994). The cognitive structure of scientific theories. *Philosophy of Science, 61*, 276–296.

Godfrey-Smith, P. (2009). Models and fictions in science. *Philosophical Studies, 143*, 101–116.

Hacking, I. (1983). *Representing and intervening: Introductory topics in the philosophy of natural science*. Cambridge: Cambridge University Press.

Hendricks, V. F., Jakobsen, A., & Pedersen, S. A. (2000). Identification of matrices in science and engineering. *Journal for General Philosophy of Science, 31*, 277–305.

Holton, G. (1978). Subelectrons, presuppositions and the Millikan-Ehrenhaft dispute. *Historical Studies in the Physical Sciences, 9*, 166–224.

Houkes, W. (2008). Designing is the construction of use plans. In P. E. Vermaas, P. Kroes, A. Light, & S. A. Moore (Eds.), *Philosophy and design: From engineering to architecture* (pp. 37–49). Dordrecht etc.: Springer.

Houkes, W. (2009). The nature of technological knowledge. In A. Meijers (Ed.), *Philosophy of technology and engineering sciences* (Handbook of the philosophy of science, Vol. 9, pp. 309–351). Amsterdam: North-Holland.

Houkes, W., & Vermaas, P. E. (2010). *Technical functions: On the use and design of artefacts*. Dordrecht etc.: Springer.

Hughes, T. P. (1983). *Networks of power: Electrification in western society 1880–1930*. Baltimore: Johns Hopkins University Press.

Jeffrey, R. C. (1956). Valuation and acceptance of scientific hypotheses. *Philosophy of Science, 23*, 237–246.

Kelly, T. (2003). Epistemic rationality as instrumental rationality: A critique. *Philosophy and Phenomenological Research, 66*, 612–640.

Kriegel, U. (2008). The dispensability of (merely) intentional objects. *Philosophical Studies, 141*, 79–95.

Kroes, P. (2009). Foundational issues of engineering design. In A. Meijers (Ed.), *Philosophy of technology and engineering sciences* (Handbook of the philosophy of science, Vol. 9, pp. 513–541). Amsterdam: North-Holland.

Lenk, H. (2007). *Global technoscience and responsibility*. Berlin etc: Lit Verlag.

Mayo, D. G. (1994). The new experimentalism, topical hypotheses, and learning from error. In D. M. Hull, M. Forbes, & R. Burian (Eds.), *PSA 1994* (Vol. 1, pp. 270–279). East Lansing: Philosophy of Science Association.

Meijers, A. (Ed.). (2009). *Philosophy of technology and engineering sciences* (Handbook of the philosophy of science, Vol. 9). Amsterdam: North-Holland.

Morrison, M., & Morgan, M. S. (1999). Models as mediating instruments. In M. S. Morgan & M. Morrison (Eds.), *Models as mediators: Perspectives on natural and social science* (pp. 10–37). Cambridge: Cambridge University Press.

Niiniluoto, I. (1993). The aim and structure of applied research. *Erkenntnis, 38*, 1–21.

Rhodes, R. (1986). *The making of the atomic bomb*. London: Simon & Schuster.

Rudner, R. (1953). The scientist qua scientist makes value judgments. *Philosophy of Science, 20*, 1–6.

Seely, B. E. (1984). The scientific mystique in engineering: Highway research at the Bureau of Public Roads, 1918–1940. *Technology and Culture, 25*, 798–831.

Simon, H. A. (1969). *The sciences of the artificial*. Cambridge, MA: MIT Press.

Stanford, P. K. (2006). *Exceeding our grasp*. Oxford/New York: Oxford University Press.

Suh, N. P. (2001). *Axiomatic design: Advances and applications*. Oxford/New York: Oxford University Press.

Thomasson, A. L. (1996). Fiction and intentionality. *Philosophy and Phenomenological Research, 56*, 277–298.

Toon, A. (2010). The ontology of theoretical modelling: Models as make-belief. *Synthese, 172*, 301–315.

van Fraassen, B. C. (2008). *Scientific representation: Paradoxes of perspective*. Oxford: Clarendon Press.

Von Wright, G. H. (1963). *Norm and action: A logical enquiry*. London: Routledge and Kegan Paul.

Wagenaar, W. A. (2001, January 18). Onderscheid wetenschap en techniek leidt ons op dwaalsporen. *Delta* (weekly magazine of Delft University of Technology), *33*(2), 13.

The manufacturer's authorised representative in the EU is Springer Nature Customer Service Centre GmbH, Europaplatz 3, 69115 Heidelberg, Germany. If you have any concerns regarding our products, please contact ProductSafety@springernature.com

Printed and bound by CPI Group (UK) Ltd, Croydon, CR0 4YY

25/03/2026

02078174-0008